Thomas MacKellar

The American Printer - A Manual of Typography

Containing Complete Instructions for Beginners, as Well as Practical Directions for

Managing Every Department of a Printing Office

Thomas MacKellar

The American Printer - A Manual of Typography
Containing Complete Instructions for Beginners, as Well as Practical Directions for Managing Every Department of a Printing Office

ISBN/EAN: 9783337252083

Printed in Europe, USA, Canada, Australia, Japan

Cover: Foto ©berggeist007 / pixelio.de

GUTENBERG AND FAUST'S FIRST PROOF FROM MOVABLE TYPES.

THE

AMERICAN PRINTER:

A Manual of Typography,

CONTAINING

COMPLETE INSTRUCTIONS FOR BEGINNERS,

AS WELL AS

Practical Directions for Managing all Departments of a Printing Office.

WITH SEVERAL USEFUL TABLES,
SCHEMES FOR IMPOSING FORMS IN EVERY VARIETY,
HINTS TO AUTHORS AND PUBLISHERS,
ETC. ETC.

BY
THOMAS MACKELLAR.

FIAT LUX.

PHILADELPHIA:
MACKELLAR, SMITHS & JORDAN.
1867.

Hereby, tongues are known, knowledge groweth, judgment increaseth, books are dispersed, the Scripture is read, stories be opened, times compared, truth discerned, falsehood detected and with finger pointed, and all (as I said) through the benefit of Printing. FOX'S MARTYRS.

Third Edition.

PREFACE.

USEFULNESS rather than originality has been aimed at in the preparation of the AMERICAN PRINTER, which is offered as an improvement on the typographical work formerly published by us. In addition to the results of actual personal experience embodied in the volume, information has been gathered from various publications, such as *Ames and Dibdin's Typographical Antiquities*, *Thomas's History of Printing*, *Timperley's Dictionary of Printers and Printing*, *Savage's Dictionary of Printing*, *Johnson's Typographia*, *Chambers's Encyclopædia*, *Beadnell's Guide to Typography*, and the other books referred to in the notes. The work has been prepared amid the manifold interruptions incident to business life; yet we think nothing has been overlooked that is essential for the instruction of the learner or for the assistance of the workman.

Besides the matter relating to practical typography, the volume contains a sketch of the discovery of printing, and notices of type-founding, stereotyping, electrotyping, and lithography. The implements employed in typography are describel and their uses explained; and complete schemes for imposition are laid down. The valuable tables and the plans of cases for various languages, and for music and labour-saving rule, will be found extremely useful; as well as the extensive lists of abbreviations and of foreign words and phrases, and orthographical hints.

Special attention has been given in setting forth the functions and duties of the foreman and proof-reader, so that the operations of an office may be prosecuted with efficiency, comfort, and economy.

Authors and publishers, as well as typographical amateurs, may consult the volume with profit; and, indeed, any intelligent person will find it a serviceable companion.

CONTENTS.

HORN-BOOK OF THE SEVENTEENTH CENTURY.

The American Printer.

The Song of the Printer.

Pick and click
Goes the type in the stick,
As the printer stands at his case;
His eyes glance quick, and his fingers pick
The type at a rapid pace;
And one by one as the letters go,
Words are piled up steady and slow—
Steady and slow,
But still they grow,
And words of fire they soon will glow;
Wonderful words, that without a sound
Traverse the earth to its utmost bound;
Words that shall make
The tyrant quake,
And the fetters of the oppress'd shall break.
Words that can crumble an army's might,
Or treble its strength in a righteous fight,
Yet the type they look but leaden and dumb,
As he puts them in place with finger and thumb;
But the printer smiles,
And his work beguiles
By chanting a song as the letters he piles,
With pick and click,
Like the world's chronometer, tick! tick! tick!

O, where is the man with such simple tools
Can govern the world as I?
With a printing press, an iron stick,
And a little leaden die,
With paper of white, and ink of black,
I support the Right, and the Wrong attack.

Say, where is he, or who may he be,
That can rival the printer's power?
To no monarchs that live the wall doth he give,—
Their sway lasts only an hour;
While the printer still grows, and God only knows
When his might shall cease to tower!

THE

AMERICAN PRINTER.

RISE AND PROGRESS OF PRINTING.

DISCOVERY OF PRINTING.

THE art which perpetuates the history and achievements of all the arts and sciences has not with certainty preserved the name of its own discoverer. At all events, the point is so dubious that several cities have advanced rival claims to the honour of the invention. We agree with ISAIAH THOMAS in the opinion that the following points may be regarded as established:—

1. That the cities of Haerlem in Holland, and Mentz and Strasburg in Germany, all claim the honour of being the birthplace of the art of printing.

2. That Laurentius, sometimes called Coster, Koster, or Kustos, has the best claim to the honour of the discovery, which was made about the year 1420, or, as several writers state, not earlier than 1422, nor later than 1436.

3. That he lived at Haerlem, was a man of large property, had a lucrative office under the government, and there practised printing in its original rude state.

4. That Laurentius, for some time after he began printing, used wooden blocks or plates, on which he engraved, or carved, in pages, &c. the words for several small works; in

some of which were pictures, cut in the blocks with the words. These he printed only on one side of vellum, or paper, and doubled and pasted the leaves together, thus forming them into books. After practising this way for a few years, he invented and used separate wooden types, but never attempted to cut or cast types in metal.*

* "About one hundred and twenty years ago, Laurence Zanssen Coster inhabited a decent and fashionable house in the city of Haerlem, situated on the market-place, opposite the royal palace. The name of Coster was assumed, and inherited from his ancestors, who had long enjoyed the honourable and lucrative office of coster or sexton to the church. This man deserves to be restored to the honour of being the first inventor of printing, of which he has been unjustly deprived by others, who have enjoyed the praises due to him alone. As he was walking in the wood contiguous to the city, which was the general custom of the richer citizens and men of leisure, in the afternoon and on holidays, he began to cut letters on the bark of the beech; with these letters he enstamped marks upon paper in a contrary direction, in the manner of a seal, until at length he formed a few lines for his own amusement and for the use of the children of his brother-in-law. This succeeding so well, he attempted greater things; and, being a man of genius and reflection, he invented, with the aid of his brother or son-in-law, Thomas Pietrison, a thicker and more adhesive ink, as the common ink was too thin and made blotted marks. With this ink he was able to print blocks and figures, to which he added letters. I have seen specimens of his printing in this manner: in the beginning he printed on one side only. This was a Dutch book, entitled *Spiegal enser Behoudenisse*. That it was one of the first books printed after the invention of the art, appears from the leaves, which are pasted together, that the naked sides might not be offensive to the eyes; and none at first were printed in a more perfect manner. As this new species of traffic attracted numerous customers, thus did the profits arising from it increase his love for the art and his diligence in the exercise of it.

"He engaged workmen, which was the source of the mischief. Among these workmen was one Jan ——: whether his surname be that of Faust, or any other, is of no great importance to me, as I will not disturb the dead, whose consciences must have smote them sufficiently while living. This Jan, who assisted at the printing-press under oath, after he had learned the art of casting the types, setting them, and other articles belonging to the art, and thought himself sufficiently instructed, having watched the opportunity, as he could not find a better, he packed up the types and the other articles on Christmas eve, while the family was engaged in celebrating the festival, and stole away with them. He first fled to Amsterdam, thence to Cologne, until he could establish himself at Mentz, as a secure place, where he might open shop and reap the fruits of his knavery. It is a known fact that within the twelve months (that is, in the year 1440) he published the *Alexandri Galli Doctrinale* (a grammar at that time in high repute), with *Petri Hispani Tractatibus Logicis*, with the same letters which Laurens had used. These were undoubtedly the first products of his press. These are the principal circumstances that I have collected from creditable persons far ad-

5. That Laurentius employed several servants in his business; among whom was John Geinsfleisch, Sr. There were two brothers of that name: the younger was sometimes distinguished by the name of Gutenberg. He was an ingenious artist, and lived at Strasburg.

6. That John Geinsfleisch, Sr., communicated, first, the theory of the art, and, afterward, the practice of it, to his younger brother; whom, for the sake of distinction, I shall hereafter call Gutenberg.

7. That Laurentius followed printing during the remainder of his life; and that, after his death, the business was continued in his family at Haerlem for many years.

8. That John Geinsfleisch, the servant of Laurentius, about the time that his master died, with the aid of a fellow-servant who was his accomplice, took an opportunity, on a festival, to steal a considerable part of his master's wooden types, with other parts of his printing apparatus, and absconded; and, having conveyed his plunder to Mentz, his native place, he there commenced printing, about the year 1440, with the types he had stolen from his master.

9. That, after Geinsfleisch settled at Mentz, he was assisted with money, &c. by John Fust, *alias* Faust, *alias* Faustus, a rich and very respectable man; who, consequently, shared the profits with Geinsfleisch. Faust and Geinsfleisch afterward formed a company, and admitted as a partner John Meidenbachius, with some other persons.

10. That Gutenberg, the younger brother of Geinsfleisch, continued at Strasburg till 1444, and was in various employments; but he made great efforts toward attaining the art of printing with cut metal types. He could not, however, bring the art to any degree of perfection. It is believed by some that he and the partners with whom he was concerned printed a few very small works. Their performances have all disappeared, and, as far as known, have been entirely destroyed. Although, whilst at Strasburg, Gutenberg had made considerable progress in improving the art, yet, having quarrelled with his partners, and being involved in lawsuits, he quitted that city and joined his brother at Mentz.

---

vanced in years, which they have transmitted like a flaming torch from hand to hand; I have also met with others who have confirmed the same," &c. &c.—*Hadrianus Junius*, 1578.

11. The two brothers had the management of the printing business at Mentz, and they united their endeavours to form a fount of metal types with cut faces. Their method of making these types was, first to cast the shanks, or bodies, to a suitable size, and afterward to engrave or cut the letters on them. After a labour of several years, they accomplished the undertaking; and in 1450 a part of the Bible appeared from their press, which was printed with those types.* The same year, and very soon after they began to work with those types, the partnership between the brothers, Faust & Company, was dissolved, and a connection between Faust and Gutenberg commenced; but, a difference between them arising, an action at law was instituted by Faust for money advanced to Gutenberg, and their joint concern in business ended in 1455. After this, Gutenberg was assisted by Conrad Humery, Syndic of Mentz, and others; and this new company opened another printing-house in that city. Faust also continued the business, and took into partnership one of his servants, called Peter Schœffer, an ingenious man, who had become very skilful in the printing business.

12. That Schœffer, in 1456, completed the invention of metallic types by casting them with faces. "He privately cut matrices for the whole alphabet; and, when he showed his master the types cast from these matrices, Faust was so much pleased that he gave Schœffer his only daughter in marriage." There were at first many difficulties with these types, as there had been with those of wood and those that were cut on metal: one was owing to the softness of the metal, which would not bear forcible pressing; but this defect, as well as some others, was soon remedied. The first book printed with the improved types was *Durandi Rationale*. It was not finished till 1459.

From these facts, it appears that the rudimental discovery of the art may be fairly allotted to Laurentius of Haerlem; the invention and improvement of metal cut-face types belong to Geinsfleisch and his brother Gutenberg; the practical and final completion of the art by the invention of metal types cast with faces is due to Peter Schœffer, of Mentz.†

---

* Polydore Virgil mentions that metal types with cut faces were first thought of in 1442.

† For a more detailed account, see Thomas's *History of Printing*, Ames's and Dibdin's *Typographical Antiquities*, *et al.*

The knowledge and practice of the art were gradually extended over Europe. A press was established at Boulogne as early as 1462; one at Paris in 1464; and another at Rome in 1466. Iceland had its printing-office in 1530, at which a Bible was printed in 1584.

William Caxton acquired a knowledge of the art in Germany, carried it into practice at Westminster in England, and in 1474 printed a book entitled *The Game of Chess.* Though at that time over sixty years old, he was remarkable for his industrious habit. He was possessed of good sense and sound judgment; steady, persevering, active, zealous, and liberal in his devices for that important art which he introduced into England, labouring not only as a printer, but as translator and author. The productions of his press amount to sixty-four. In the churchwardens' books of St. Margaret's Parish, Westminster, his death is thus recorded:—"1491. Item, atte bureyng of William Caxton, for iiii. torches vj*s.* viij*d.* Item, for the belle atte same bureyng, vj*d.*"

ANCIENT PECULIARITIES.

THE pages were either large or small folios, but sometimes quartos; the smaller sizes were not in use.

The leaves were without running title, direction-word, number of pages, or divisions into paragraphs.

The character itself was a rude old Gothic mixed with Secretary, designed to imitate the handwriting of the times; the words were printed so close to one another that the matter was difficult and tedious to be read, even by those who were used to manuscripts and to this method, and often led the inattentive reader into mistakes.

Ancient printers did not divide words at the ends of lines by hyphens. To avoid divisions, they used vowels with a mark of abbreviation to denote that one or more letters were omitted in the word: *e.g.* cōpose for compose, cōpletiō for completion, &c. No punctuation-marks were used, except the colon and full point; but an oblique stroke (/) was after a while introduced, for which the comma was finally substituted.

Orthography was various and often arbitrary, and syntax was disregarded. Proper names and sentences were often begun with small letters, as well as the first words in lines of poetry.

Blanks were left for the places of titles, initial letters, and other ornaments, to be supplied afterward by illuminators, whose ingenious art, though in vogue before and at that time, did not long survive the masterly improvements made by the printers in this branch of their art. These ornaments were exquisitely fine, and curiously variegated with the most beautiful colours, and even with gold and silver; the margins, likewise, were frequently charged with a variety of figures, of saints, birds, beasts, monsters, flowers, &c., which had sometimes relation to the contents of the page, though often none at all. These embellishments were often very costly.

The name of the printer, place of his residence, &c. were either wholly neglected or put at the end of the book, not without some pious ejaculation or doxology.

The date was likewise omitted, or involved in some cramped, circumstantial design, or printed either at full length or in numerical letters, and sometimes partly one and partly the other: thus, One Thousand CCCC and lxxiiii; but always at the end of the book.

There was no variety of character, nor intermixture of Roman and Italic,—which were of later invention; but the pages were all printed in a Gothic letter of the same size throughout.

About 1469-70, alphabetical tables of the first words of each chapter were introduced, as a guide to the binder. Catchwords (now generally abolished) were first used at Venice, by Vindeline de Spire. The inventor of signatures is unknown.

PRINTING IN AMERICA.

THE first press introduced into America was at Lima, Peru, about the year 1590, and the next into Mexico, in 1600.

Cambridge, Massachusetts, is entitled to the distinction of having the first printing-press in North America, which was under the charge of Stephen Daye. For this press the colony was mainly indebted to the Rev. Jesse Glover, a nonconformist minister possessed of a considerable estate, who had left England to settle among his friends in Massachusetts. Some gentlemen of Amsterdam also "gave towards furnishing of a printing-press with letters, forty-nine pounds and something more." This was about 1638. The first book issued was the *Bay Psalm-Book*, in 1640.

The first book issued in the Middle Colonies was an Almanac, printed by William Bradford in 1685, near Philadelphia.* Bradford was brought out from England in 1682 by William Penn, who was desirous to give his prospective colony the benefit of a printing-press. As the government of Pennsylvania became very restrictive in regard to the press, Bradford in 1693 removed to New York, and was appointed printer to that colony, where, in connection with his business, he established in 1725 the *New York Gazette*, the first newspaper published there. He died May 23, 1752, after an active and useful life of eighty-nine years.

The first newspaper in America was the *Boston News Letter*, which was first issued by John Campbell on Monday, April 24, 1704: it was regularly published for nearly seventy-two years. The second was the *Boston Gazette*, begun December 21, 1719. The third was the *American Weekly Mercury*, issued in Philadelphia on December 22, 1719. James Franklin, an elder brother of Benjamin, established the *New England Courant*, August 17, 1721.

TYPE-FOUNDING IN EUROPE.

For a long period after the discovery of printing, it seems that type-founding, printing, and binding went under the general term of *printing*, and that printers cast the types used

---

* Two copies of this Almanac are known to be in existence. We give the Address of

THE PRINTER TO THE READERS.

Hereby underſtand that after great charge & Trouble, I have brought that GREAT ART & MYSTERY OF PRINTING into this part of America; believing it may be of great ſervice to you in ſeveral reſpects; hoping to find encouragement, not only in this Almanack, but what elſe I ſhall enter upon for the uſe & ſervice of the Inhabitants of theſe Parts. Some irregularities there be in this Diary, which I deſire you to paſs by this year; for being lately come hither, my materials were miſplaced & out of order, whereupon I was forced to uſe Figures & Letters of various Sizes: but underſtanding the want of ſomething of this nature, & being importuned thereto, I ventured to make public this; deſiring you to accept thereof; & by the next (as I find encouragement) ſhall endeavour to have things compleat. And for the eaſe of Clarks, Scriveniers, &c., I propoſe to print blank Bills, Bonds, Letters of Attorney, Indentures, Warrants, etc., & what elſe preſents itſelf, wherein I ſhall be ready to ſerve you; and remain your friend.

W. BRADFORD.

Philadelphia, the
10th month, 1685.

by them, and printed and bound the works executed in their establishments. Type-founding became a distinct calling early in the seventeenth century. A decree of the Star Chamber, made July 11, 1637, ordained the following regulations concerning English founders:—

"That there shall be four founders of letters for printing, and no more.

"That the Archbishop of Canterbury, or the Bishop of London, with six other high commissioners, shall supply the places of those four as they shall become void.

"That no master-founder shall keep above two apprentices at one time.

"That all journeyman-founders be employed by the masters of the trade, and that idle journeymen be compelled to work, upon pain of imprisonment and such other punishment as the court shall think fit.

"That no master-founder of letters shall employ any other person in any work belonging to the casting or founding of letters than freemen or apprentices to the trade, save only in pulling off the knots of metal hanging at the ends of the letters when they are first cast; in which work every master-founder may employ one boy only, not bound to the trade."

By the same decree, the number of master-printers in England was limited to twenty.

Regulations like the above were in force till 1690. The "polyglot founders," as they have been called, were succeeded by Joseph Moxon and others. But the English were unable to compete with the superior productions of the Dutch founders, until the advent of William Caslon, who, by the beauty and excellence of his type, surpassed his Batavian competitors, when the importation of foreign type ceased, and his founts were, in turn, exported to the Continent.

TYPE-FOUNDING IN AMERICA.

A FOUNDRY, principally for German type, was established at Germantown, Pennsylvania, about the year 1735, by Christopher Saur (or Sower), a printer, who executed in German the first quarto Bible printed in America, as well as other valuable works in the German language. Three editions were printed of the Bible,—viz. in the years 1743, 1763, and 1776, the latter two by his son. In 1739, Saur published a newspaper in Germantown.

An abortive attempt was made about 1768 to set up a foundry at Boston, by a Mr. Mitchelson, from Scotland, and another in Connecticut, by a Mr. Buck. In 1775, Dr. Franklin brought from Europe to Philadelphia the materials for a foundry; but little use was made of them.

John Baine, a type-founder of Edinburgh, sent a relative to this country with tools for a foundry at the close of the Revolutionary War, and soon after came over himself. They carried on the business till 1790, when Mr. Baine died, and his kinsman returned to Scotland.

A Dutch founder afterward settled at New York, and cast Dutch and German faces, which were considered handsome, though his Roman styles were very poor. Want of money prevented his success.*

In 1796, type-founding was commenced in Philadelphia, by Archibald Binny and James Ronaldson, natives of the city of Edinburgh, where Binny had carried on the same business. Their assortment was not extensive, but it embraced the essential founts,—Brevier, Bourgeois, Long Primer, Small Pica, Pica, and two-line letters. They were obliging and attentive, and in twenty years made a fortune.† They improved their foundry according to the increase of printing and the consequent demands of the trade, extending their assortment from Pearl, of 180 lines in a foot, to 12-line Pica, having 6 lines. Binny made an important improvement in the type-

* For the remainder of this article we are largely indebted to the venerable George Bruce, of New York.

† After the retirement of Binny & Ronaldson, Richard Ronaldson carried on the business of this foundry until 1833, when he in turn was succeeded by Lawrence Johnson and George F. Smith. Mr. Johnson, a man of energy and enterprise, had introduced stereotyping into Philadelphia, and now both callings were incorporated. Ten years afterward, Mr. Smith retired; and in 1845 Mr. Johnson associated with him Thomas MacKellar, John F. Smith, and Richard Smith, who had, as it were, grown up with the business. The foundry now quickly grew in importance, and won a wide reputation. A quarterly periodical, entitled the *Typographic Advertiser*, edited by a then junior of the firm, was (and still is) issued and circulated at home and in foreign lands; while its finely-printed Specimen Books showed that the foundry was making rapid progress. Mr. Johnson died in 1860, and was succeeded by his three junior partners, who, with Peter A. Jordan, constitute the present firm, now known as MacKellar, Smiths & Jordan, under whose management the establishment has been brought to rank at least equal with the first in the world. Their new Specimen Book may be regarded as a unique mechanical and literary production.

mould, by which a caster could cast 6000 letters in a day with as much ease as he before could cast 4000.

According to Holmes's *American Annals*, about 200 newspapers were printed in the United States in the year 1801, of which 17 were issued daily, 7 three times a week, 30 twice a week, and 146 weekly. There must also have been at the same time as many as 60 offices engaged in miscellaneous printing. The whole business had increased threefold in eleven years. Another type-foundry was put in successful operation in Baltimore, about 1805, by Samuel Sower & Co. It had in it some moulds and matrices which had been used by Christopher Sower, who had printed in Germantown, near Philadelphia, and cast his own types. He printed with German characters; but now the foundry was revived with excellent Roman and Italic letters, and among other extraordinary things it had the size called Diamond, with a smaller face than had ever been cast before. It was the smallest type in the world.

The demand for type was very brisk till the war of 1812 commenced, and the foundries were generally three or four months in arrears in their execution of orders. The names of the newspapers published in the United States in April, 1810, are given in Thomas's *History of Printing*, and amount to 359, of which 27 were daily papers, 38 were printed twice, 15 three times, and 279 once in a week. Add those required for general printing, and the whole number of offices could not be less than 500,—being an increase of 240 in nine years, and some of them using several thousand pounds of type for book-printing.

In 1811, Elihu White established a type-foundry in New York. He had been long engaged, in connection with Mr. Wing, in the manufacture of printing-types, at Hartford, Connecticut, upon a plan of their own invention, by which 20 or 30 letters were cast at once; but, abandoning that invention, he adopted the old plan of casting, and, having a good assortment of faces and bodies, his removal to New York was a great convenience to its printers, and they gave him a very satisfactory support. But the principal business in type-founding still continued for some years to be done in Philadelphia.

In 1813, another type-foundry was begun in the city of New

York, by D. & G. Bruce, principally to cast types for their own use. They had carried on book-printing for seven years, and had now become acquainted with the stereotype art,—Mr. David Bruce having visited England in 1812 and acquired it by purchase and actual labour. For ordinary printing, it was customary to bevel off the body of the type at the face end, or shoulder, as it is usually called, which unfitted it for making a strong stereotype plate in the most approved way: hence the necessity for casting type expressly for stereotype. Their first fount was Bourgeois, with which they cast two sets of plates of the New Testament (the Common School Testament), and sold one of these to Mathew Carey, of Philadelphia, retaining the other for their own business. But these were not completed till 1814. In 1815, they cast the plates of the 12mo School Bible, on Nonpareil type, prepared, like the Bourgeois, at their own foundry expressly for stereotyping. They thus gave the first stereotype School Testament and School Bible to America; but not the first stereotype book. John Watts, of England, also commenced stereotyping in New York in 1813, and completed the Westminster Catechism that year, a volume of 120 pages 12mo. David Bruce invented the planing-machine for equalizing the thickness of stereotype plates, which is now used in every stereotype foundry in the United States. The process of stereotyping is, however, entirely different from that of ordinary type-founding, and it is, therefore, generally carried on as a separate business, or connected with the composing department of a printing-office. Twenty compositors and two proof-readers will furnish full employment for one moulder, one caster, and three finishers, who will, among them, complete, on an average, 50 pages of octavo per day.

In 1818, or soon after, a type and stereotype foundry was established in Boston, and another in Cincinnati, principally through the enterprise of the late Elihu White, who, having the means of multiplying matrices with facility, took this method for the extension of his business. Others followed his example, and type-foundries were established in Albany, Buffalo, Pittsburgh, Louisville, and St. Louis, with several additional in New York, Boston, Philadelphia, and Baltimore. The business, in fact, was overdone, and failures and suppressions took place, as competition reduced the prices of types.

The mode of type-founding has latterly undergone some important changes, which must no doubt be considered improvements. First among them is the introduction of machine-casting, in which a pump forces the fluid metal into the mould and matrix, and gives a sharper outline to the letter than was formerly given by the most violent throw of the caster. The old practice of casting only a single type at a time remains. The first idea of this machine originated with William M. Johnson, who obtained a patent for it in 1828. Elihu White put it into use in his type-foundry, and persevered in using and trying to improve it as long as he lived; but he did not succeed in removing the greatest fault, which was a hollowness in the body of the type cast by it, that inclined them to sink under the pressure of the printing-press. The first successful type-casting machine was invented by David Bruce, Jr., of New York, and was patented March 17, 1838. The patent was sold to George Bruce, and the machines were used by him until 1845. David Bruce meanwhile patented another machine in 1843, which, with new improvements, patented two years later, gave entire satisfaction, and is now in general use in American foundries. By Bruce's machine, three times the quantity of type that was cast by Binny & Ronaldson's improved mould is now cast in a given time, and nearly five times the quantity that was cast by the common hand-mould fifty years ago. This improvement has passed into Europe, and been adopted by most of the German type-founders; but in Britain for some time it found little favour; and some kinds of types are still cast there in the same kind of mould as was used two hundred years ago, or in the earliest-known type-founding, at the rate of 4000 letters in a day.

The next innovation to be mentioned is the application of electrotyping in making copies of new styles got up by rival or foreign founders. In regular type-founding, the original of each character is formed on a separate steel punch, which, being hardened and tempered, is driven into copper a sixteenth of an inch or more, to form the face of the type, called a matrix. The matrices thus made are more durable than those produced by galvanism; the latter being formed by precipitating copper on the face of a type in a galvanic battery.

There are now three type-foundries in Boston, eight in the city of New York, one in Buffalo, three in Philadelphia, one in Baltimore, two in Cincinnati, one in Chicago, one in Milwaukee, and one in St. Louis,—in all, twenty-one. These foundries not only supply the printers of the United States, but most of the printers in Canada, some in the British West India Islands, the Spanish and Danish Islands, Mexico, and South America. The quality of American types is equal—if not superior—to that of any made in Europe. The following are the prices at which they have been sold for the last sixty-six years, given at nine different dates, and naming only the principal and most useful sizes.

	1801.	1806.	1811.	1819.	1827.	1831.	1841.	1850.	1866.
Pica	$0.35	$0.44	$0.55	$0.44	$0.42	$0.36	$0.38	$0.30	$0.56
Small Pica	.40	.48	.58	.48	.46	.38	.40	.32	.58
Long Primer	.47	.56	.66	.56	.50	.40	.42	.34	.62
Bourgeois	.56	.66	.76	.66	.58	.46	.46	.37	.66
Brevier	.67	.76	.86	.76	.70	.56	.54	.42	.70
Minion		1.03	1.13	1.00	.88	.70	.66	.48	.76
Nonpareil	1.12	1.40	1.75	1.40	1.20	.90	.84	.58	.84
Agate					1.44	1.10	1.08	.72	1.00
Pearl					1.75	1.40	1.40	1.08	1.40
Diamond								1.60	1.80

STEREOTYPING.

STEREOTYPING is said to have been invented by J. Van der Mey, in Holland, about 1698. A quarto Bible and some other books were printed by him from plates, which were formed by soldering the bottoms of common type together. William Ged, of Edinburgh, discovered the present mode, and stereotyped the Bible and Book of Common Prayer in 1725. He encountered malicious opposition, and the business was abandoned, the new method dying with the inventor. About 1745, Benjamin Mecom, a nephew of Dr. Franklin, cast plates for a number of the pages of the New Testament. Dr. Alexander Tilloch, of Glasgow, re-discovered the art in 1781. Stereotyping gradually spread, and became a potent means in the reduction of the cost of books. At present, almost every important work is stereotyped or electrotyped.

Several methods of stereotyping are now practised. Many of the leading newspapers of England and America are printed from stereotype plates cast in moulds made of prepared paper: this mode is, however, very inferior, and is not applicable to fine books.

Matter for stereotyping is set with high spaces and quadrates. The forms must be small, containing about two pages of common octavo. A slug type-high is put above the top line and another below the foot line of each page, to protect the ends of the plates from injury when they are passed through the shaving-machine. Bevelled slugs, in height equal to the shoulder of the type, are placed on both sides and between the pages, to form the flange by which the plate is to be clasped by the hooks of the printing-block.

Before the form is sent into the foundry, the type must be carefully compared with the proof, to detect any errors which may have been left uncorrected. Care must be taken to lock up the form perfectly square and quite tight, to prevent the types from being pulled out when the mould is raised from the pages. It must be evenly planed down, and no ink or dirt or incrustations from the ley be allowed to remain on the surface.

The face of the type being clean and dry, and the bottoms free from particles of dirt, the form is laid on a clean moulding-stone, and brushed over with sweet-oil, which must be laid on as thinly as possible, care being taken that the entire surface of the types is covered. A moulding-frame, with a screw at each corner (called a *flask*), and fitting neatly to the form, is next placed around it.

The material for moulding is finely ground gypsum, nine parts of which are mixed with about seven parts of water, and well stirred up. A small quantity of the liquid mixture is poured over the pages, and gently pressed into the counter of the types with a small roller, for the purpose of expelling confined air; after which, the remainder of the gypsum is poured in, until the mould is somewhat higher than the upper edge of the flask. In a few minutes the mixture sets, and the upper side is smoothed over with a steel straight-edge. In about ten minutes the mould is gently raised by means of the screws at the corners of the flask; and, after being nicely trimmed at the sides, and nicked on the surface-edges to make openings for the metal to run in, it is placed on a shelf in an oven, and allowed to remain until the moisture has quite evaporated.

The casting-pans may be large enough to hold three or four moulds. The dried moulds are placed in a pan face downward, upon a moveable iron plate called a floater. The cover of the casting-pan, which has a hole at each corner for the passage of the metal, is then clamped to it, and lifted by a movable crane and gently lowered into the metal-pot,—containing, it may be, a thousand pounds of liquid metal,—till the metal begins to flow slowly in at the corners. When the pan is filled, it is sunk to the bottom of the pot. The metal should be hot enough to light a piece of brown paper held in it. After being immersed eight or ten minutes, the pan is steadily drawn out by means of the crane, and swung over to the cooling-trough, into which it is lowered and rested upon a stono so as just to touch the water, in order that the metal at the bottom of the pot may cool first. The metal contracts while cooling, and the caster occasionally pours in a small quantity at the corners from a ladle, till it will take no more.

The plates are carefully removed from the solid mass which comes out of the pan, and the plaster is washed from the surface. If, after examination, the face is good and sharply set, the plates are passed over to a picker, who removes any slight defects arising from an imperfection of the mould. They are then trimmed and passed through the shaving-machine, till all are brought to an equal thickness. The flanges are neatly side-planed, and the plates are then boxed, ready for the printing-press.

In England, the plates are merely turned on the back, and, consequently, vary in thickness. This must be a source of continual expense and annoyance to the pressman. The flanges, besides, are very imperfectly made,—so imperfectly that they cannot be used on American printing-blocks; and English plates, when imported into this country, are, therefore, sent to a foundry here, to be brought to an equal thickness and to be properly side-planed. The American shaving-machine and printing-block are scarcely known abroad, though far superior to foreign arrangements.

In 1804, before the introduction of stereotyping into this country, Mathew Carey, the well-known enterprising publisher in Philadelphia, had the Bible in quarto set up entire, and regularly imposed in chases, to print from at convenience, according to the demand for the volume. The type was cast by Binny & Ronaldson. Stereotyping would have saved a vast proportion of the immense outlay required to carry out the scheme, which, nevertheless, even under these circumstances, was doubtless highly remunerative. The weight of type must have amounted to 25,000 pounds, to say nothing of the number of chases and column-rules required.

ELECTROTYPING.

LATELY, stereotyping has measurably given way to the application of galvanism named Electrotyping, and may be superseded by it, though the former is better for moderate editions.

The pages being made ready and laid in a press, a pan of prepared wax, warmed, is placed over the pages and pressed down to the counter of the types. The mould is carefully dusted with plumbago, to give it a metallic surface, and is then suspended in the battery. On this, in a few hours, is deposited a thin shell of copper, which, after being coated with tin solder, is backed up with metal to the usual thickness of a stereotype plate. This method of electro-stereotyping is desirable for Bibles and other works of which immense editions are required. Wood-cuts are usually electrotyped, as a stereotype mould cannot be drawn unless the wood-cut has been previously coated with gum, which thickens the lines and injures the effect of the engraving.

The same care in preparing the pages for electrotyping must be observed as for stereotyping. For stereotyping, high slugs are placed only at the top and foot of the page; but, for electrotyping, they must be set around on all sides, and the bevelled flange must be made by side-planing.

LITHOGRAPHY.

LITHOGRAPHY is the art of printing, by a chemical process, from designs made with a greasy material upon stone. It was discovered about the beginning of the present century by Alois Senefelder, an actor of Munich, Bavaria, whose patience and perseverance under the most disadvantageous circumstances were truly remarkable and praiseworthy. Differing from all other methods of printing, the impressions are obtained (by strict attention to chemical affinity) from a level surface.

The stone best calculated for lithographic purposes is a sort of calcareous slate found on the banks of the Danube, in Bavaria, the finest being found near Munich. A good stone is porous, yet brittle, of a pale yellowish drab, and sometimes of a gray neutral tint. The stones are formed into slabs from one and a-half to three inches in thickness. To prepare them for use, two stones are placed face to face with some fine sifted sand between them, and then are rubbed together with a circular motion, to produce the requisite granulation, which is made finer or coarser to suit the purpose of the artist.

The principal agents used for making designs on stone are called lithographic chalk and lithographic ink. They are composed of tallow, virgin wax, hard tallow soap, shellac, sometimes a little mastic or copal, and enough lampblack to impart a colour to the mass. These ingredients are put into an iron sauce-pan, and exposed to a strong fire till the mass is in a state of ignition. When the quantity is reduced one-half, the pan is carefully covered, or put into water to extinguish the flame and cool the mixture. After being well worked up, it is formed into small cakes or sticks. The ingredients are the same in the chalk and the ink, but the proportions are varied, and a little Venice turpentine is often added to the latter. The chalk is used in a dry state; but the ink is dissolved by rubbing in water, and is used in a pen or with a camel-hair pencil. The presence of soap renders it soluble in water.

The artist completes a drawing with the chalk upon a grained stone as he would make a drawing in pencil or chalk upon paper. If while in this state a wet sponge were passed over the face of the stone, the drawing would wash off. To prevent this, and to make it capable of yielding impressions, a weak solution of nitrous acid is poured over it, which unites with and neutralizes the alkali or soap contained in the chalk, and renders it insoluble in water. After this, the usual course is to float a solution of gum over the whole face of the stone; and, when this is taken off, the drawing is no longer removable by the application of a wet sponge, because the chalk is now insoluble. The stone is now ready for the printer, who obtains impressions by the following process.

Having damped the surface of the stone equally with a sponge filled with water which has been slightly tinctured by acid, the printer finds that the water has been imbibed by only those parts of the stone which are not occupied by the drawing, which, being greasy, repels the water and remains dry. A roller covered with ink is now passed over the stone, which will not even be soiled where it is wet, from the antipathy of oil and water. But the parts occupied by the drawing, being dry and greasy, have an affinity for the printing-ink, which, therefore, leaves the roller and attaches itself to the drawing. In this state it is said to be charged or rolled in. A sheet of damped paper is then put over it, and, the whole being passed through a press, the printing-ink is transferred from the stone to the paper, and the impression is obtained. Great nicety is requisite in the preparation of all the agents employed in this art, and in the process of printing as well as in making the drawing on the stone.

The most important application of this process is in the production of copies of coloured drawings and paintings,—a process known as *chromo-lithography*. The object here being to produce as nearly as possible fac-similes in colour, touch, and texture, as well as in drawing and light and shadow, of pictures from the pencils of painters of the highest standing, it has been found necessary to employ a large number of stones, in order to produce the almost infinite varieties of tints which are found united in a single picture,—every stone giving a separate impression in its own particular colour or tint. The mode of procedure is somewhat as follows. First, an outline of the entire subject is made by means of transfer paper, or otherwise, on a stone which is called the outline or keystone of the work. This stone yields impressions which are transferred as guides to all the other stones. On a second and third stone which serve as the basis of the print the general effect of the drawing is washed in, and from these are printed what may be called the chiaroscuro, in a faint tint of sepia and of a neutral colour or gray,—corresponding, in fact, very nearly to the neutral or dead colouring of a water-colour drawing in the method adopted by the early water-colour painters. The stones which follow are each charged with a particular colour or tint, and each leaves its impression on only a particular portion of the print,—one stone printing only the parts which are intended to be yellow or a modification of yellow, another red, another blue, and so on. Other stones charged in parts with grays or secondary colours serve to blend and harmonize the crude colours; others follow which modify these; and, finally, one gives the sharp dark touches, and is usually followed by another which supplies a sort of glaze or finishing wash, and subdues and harmonizes the whole. Of course, we have merely indicated the general method. It will be understood that the sequence of the colours in the printing, the special quality and strength to be given to each particular tint, the effect to be produced by their superposition, and many other particulars, have all to be taken into account in planning the arrangement of the colours on the stones;—since a sequence in some respects different, and an entirely different modification of colours, have to be employed for the works of most artists; and it happens that much of the colour on each of the earlier stones is covered by that of succeeding stones, and that thus only can the broken tints of the original be imitated. It is, in fact, only by watching the progress of a print through all its stages that any clear idea can be obtained of the beauty and accuracy of the whole process, of the prevision that must be exercised, and of the skill, care, and taste required at every step to carry it to a successful termination. For some of the more elaborate prints, from thirty to forty stones have been required to produce a finished print. And in order to produce this print, it must be borne in mind that each sheet of paper has to be passed as many times through the press as there are stones, since each stone imprints upon it only its own particular section of the work. Of course, in proportion to the increase in the number of the stones, does the difficulty increase of making the work upon each fall exactly upon its proper place in the general design; for, if any one were misplaced only the fiftieth of an inch, the drawing and colour of the whole would be disturbed. Hence it is found necessary to arrange the *register*, or adjustment of the stones, with the utmost care and precision, and to exercise the most careful supervision in the printing (which is entirely a hand process), since the sheet of paper expands considerably in passing through the press, and has to be dried and re-damped before it can be passed through again. But practically this is all accomplished with seeming ease, and a large and most complex subject will be found, when the last stage has been reached, to bear the most minute scrutiny; and the result, even when the copy is placed alongside the original, will surprise and delight equally those who have followed the work through its several steps, and those who may only examine the completed work.

ENGRAVING.

THE origin of engraving on wood, like that of many other useful arts, is obscured by clouds which the learned have in vain endeavoured to dispel. The most probable opinion is that it is of Asiatic origin. China seems to have the best claim to the invention. The earliest specimen of engraving on wood in Europe is supposed to have been executed about the year 1284.

According to Vasari, the important discovery of chalcography was made by Thomas Finiguerra, a Florentine goldsmith of the fifteenth century, who lived from 1400 to 1460. The manner in which he made this discovery is thus stated by the Rev. T. F. Dibdin:—

"Of engraving upon copper, the earliest known impression is that executed by one Tommaso Finiguerra, a goldsmith of Florence, with the date of 1460 upon it. One of the following circumstances is supposed to have given rise to the discovery. Finiguerra chanced to cast, or let fall, a piece of copper, engraved and filled with ink, into melted sulphur; and, observing that the exact impression of his work was left on the sulphur, he repeated the experiment on moistened paper, rolling it gently with a roller. This origin has been admitted by Lord Walpole and Mr. Landseer; but another has been also mentioned by Huber. 'It is reported,' says he, 'that a washerwoman left some linen upon a plate or dish on which Finiguerra had just been engraving, and that an impression of the subject engraved, however imperfect, came off upon the linen, occasioned by its weight and moistness.'"

As a fitting conclusion to this part, we copy an article, originally published in L. Johnson & Co.'s *Typographic Advertiser*, entitled

A WALK OVER OUR FOUNDRY.

Mr. Typograph, how are you, sir? Glad to see you. How is business with you? Plenty to do, and customers paying up? You are so prompt in paying us, that we have no doubt you have a noble set of customers. You wish to add to your stock our new things? All right, sir. You have a fine office already, but you want to keep up with the times, and give your patrons the best the type-founder can invent? That's the way, sir. The man on the lookout sees the sun the earliest.

Mr. Faithful, show our new things to Mr. Typograph, and take his order.

You say, Mr. Typograph, that you have never gone over a type-foundry? We shall be happy to show you every thing. This way, sir. Here is the metal-house. These piles of dull

METAL-HOUSE.

lead, these casks of sparkling antimony, this copper, and this tin, go to form the grand amalgam of which type is made. The worthy and kind-hearted man who is stirring at the kettle, unites, in bonds stronger than matrimony, some ten to twelve thousand pounds of these metals every week. It may appear to you, Mr. Typograph, to be a simple thing to throw into the kettle certain amounts of lead and antimony, and copper and tin, and produce type-metal. Not so, good friend. It is not an easy matter to compose a metal that shall be hard, yet not brittle; ductile, yet tough; flowing freely, yet hardening quickly. All these conditions must be—and are—met. Break a bar in two, and examine the grain of our metal: is it not beautiful?

Now, sir, let us up-stairs and see how these bars are fitted for printer's use. This is a punch-cutter—a man of exquisite finger and unerring eye—sitting amid keen and delicate tools and accurate gauges. On the end of a piece of steel, he is forming a letter. A touch here and a touch there, and frequent testing by gauges,—so he proceeds, till the letter is done; then another, and another, till the alphabet is complete; all the letters harmonizing entirely in height, breadth, appearance, length of stroke, &c. A smoke-proof of the dies is taken, and if approved the dies are one by one placed in a stamping machine, so,—and an oblong piece of copper is set under it, so,—and then this lever is brought down, so,—and a perfect impression of the die is left, as you see, deep in the copper. This is the matrix. The matrices are passed over to other workmen in the adjoining room. Observe now the carefulness and skill exercised in fitting up these bits of copper, so that, when placed in the mould, the types cast in them shall range accurately and be of uniform height. The slightest variation would give the zigzag appearance which you may have noticed in badly-made type. This we endeavour sedulously to avoid, and with how much success you can judge from our Specimen Book. Look at this drawer full of matrices. You say they are triumphs of art? True saying, evincive of good judgment.

PUNCH.

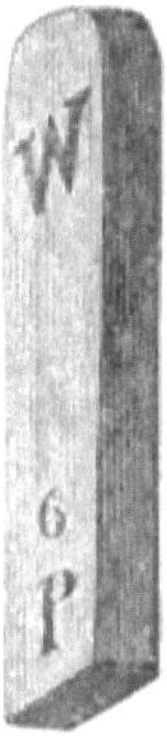

MATRIX.

You wonder what these curious-looking instruments are which lie, in dusty repose, on the shelves around the room? Those, Mr. Typograph, are hand-moulds, and at one time they provoked intense covetousness on the part of rival founders. One of our earliest predecessors, Mr. Archibald Binny (our foundry dates from 1796), added such valuable improvements to the ordinary mould, that no other foundry in the world could rival the expedition and accuracy with which types were cast in the establishment of which he was a co-proprietor. Their day has passed, however. They have been superseded by the machines which you will see in operation in another apartment. But they were capital things in their time, sir, and we regard them with somewhat of an antiquary's reverence.

Now we enter the casting-rooms. These tiny machines,

CASTING MACHINE.

small as they are, can throw out more type in one day than you would be likely to count in a month, even if you could call off one hundred a minute, and occupy ten hours a day. Snug little fellows, are they not? They were invented by a New-Yorker, Mr. David Bruce, Jr. A very ingenious man, you say? That is true. Look at one carefully. The metal is kept fluid by a little furnace underneath, and is projected into the mould by a pump, the spout of which, you see, is in front of the metal-pot. The mould is movable, and at every revolution of the crank in the hand of the workman it comes up to the spout, receives a charge of metal, and flies back with a fully-formed type in its bosom; the upper half of the mould lifts, and out jumps a type as lively as a tadpole. You don't see how the letter is formed on the end of the type? True, we had forgotten: well, this spring in front holds in loving proximity to the mould a copper matrix, such as you saw just now in the fitting-room. The letter a, for instance, stamped in the matrix, sits directly opposite the aperture in the mould which meets the spout of the pump; and when a due proportion of a's is cast, another matrix with b stamped in it takes its place; and so on throughout the alphabet. Slow work, you say, one at a time? Well, the world is peopled after that fashion; and it fills up fast enough. But just time this machine: it is making small, thin type. Count the type made in a minute. One hundred and seventy-five, you say. One hundred per minute will probably be the average of the ordinary sizes of printing type.

The types are not finished yet? Oh, no. These nimble-fingered boys are breaking off the jets, or waste ends of the type. Quick, a'n't they? Now let us go up-stairs into the dressing-room. An immense beehive? Yes, indeed, it looks like one. The lads clustered around the large circular stones, with leather-protected fingers, rub off the rough edges of the type. But men as well as type require their rough edges taken off before they are good for much in the world. These boys at the tables set up the type in long lines. You think that if you could pick up dollars as fast as they pick up type, you would retire an independent man in a year or two? We wish you could, Mr. Typograph; we wish you could.

The lines of type now pass into the hands of the dresser. Observe how deftly he slips them into a long stick, shakes them down on their face, screws them up, fastens them into a planing-board, and with one or two pushes with a planing tool accurately grooves the bottom of the type, removing entirely the burr left when the jet is broken off, and giving each type a pair of legs to stand upon, till it is worn out and returned to the melting kettle. What is the eye-glass used for? Why, sir, as soon as the types are grooved, the dresser narrowly inspects the face of the type, and if an imperfect letter is discovered by the aid of the magnifying glass, it is incontinently turned out. Ah, sir, if we were all inspected as severely as he criticizes type, some of us, perhaps, would hardly pass muster. The immaculate types are next put up in pages of convenient size, and are ready for the purchaser.

Let us drop into this side-room. Here we fit up our machines, make our moulds, repair damages to machinery, &c. The multifarious uses of these lathes you must be familiar with: this ponderous machine is an iron-planer: how it makes the iron chips fly! What is that curiously-arranged lathe? That is for cutting Labour-Saving Rule,—the rule which you have found so convenient and economical in your job-room. We make it now of nine different styles of face: one single, two dotted or hyphen-lines, two parallel, and four double, of varying thickness. They are all cut to Pica ems in length, and are furnished with mitred corner-pieces, so contrived, in the case of the three larger sizes, as to allow the rule to be used single or double, and with the fine lines inside or outside. Our brass is carefully rolled by the best manufacturers in the country, and is sent to us in sheets. That wicked-looking shears yonder cuts up the thinner sheets of brass with as much unction as Commissioner Yeh's executioner slices off heads: the thick brass goes under a circular steam-saw.

Now, sir, while we are up here, we will peep into the printers' furnishing-room. Isn't this a beautiful stereotype-block? Doesn't it do your eyes good to look at it? And these brass galleys, and mahogany galleys and composing-sticks, are they not admirable? Our effort in this department, as in all others, is to do our work well. All our miscellaneous wood-work is done here,—stands, racks, drawers, stereotype and packing boxes, &c. Some curious work has been designed and executed for the Smithsonian Institution, as well as brass ciphering-frames for the blind.

Ah, we had forgot to show you our large-type room. On

our way to the electrotype department, we will glance in it. The types you see here cool too slowly to be cast in a machine, so we continue to pour them. Look over the drawers, and see the multitude of patterns. Some men fancy one style, and some another. So we try to meet all tastes. Feel how solid the type is. You can't squeeze the life out of that type on a power-press. No, indeed. It is made for wear.

Now, Mr. Typograph, we enter the grimed and murky electrotype-room. Electrotyping, you are aware, is simply stereotyping in copper. Its advantages over ordinary stereotyping are, sharpness of outline in wood-cuts, and great durability. We electrotype a book occasionally; but the art is mainly applied to the production of duplicates of cuts, jobs, binders' stamps, &c. The thing to be electrotyped is laid upon a press, and a prepared mould is placed over it, and an exact impression taken. This is well dusted with plumbago, and then deposited in a galvanic battery. Nature immediately takes up her part of the work, and a brilliant coating of copper is deposited upon the mould. When sufficiently thick, it is taken out of the battery, and, as you may notice, presents on the wrong side the appearance of a printed sheet of copper. This sheet is then filled up on the back to the requisite degree of thickness, and fastened to a block, ready to be used with type on a common printing-press. Plumbago, you remark, does not improve the countenances of the operatives? True; but a little soap and water, vigorously applied, proves the title of these intelligent workmen to rank among white folks. The gas that chokes you is from the batteries; and so, if you please, we will proceed to the stereotype department of our business.

To you, Mr. Typograph, our composing-rooms present nothing new, except, perhaps, in the enormous size of our founts of plain type, and the great number of jobbing founts. So we will only say, that in ten years we have set up in these rooms and stereotyped more than eight hundred considerable works,—most of them consisting of a single volume, but some of from two to twelve volumes each,—besides a multitude of smaller books, tracts, &c. Among the rest we may mention two Quarto Bibles (one of them, published by Peck & Bliss, the grandest ever got up in America), Lippincott's two great Gazetteers, Dr. Kane's Explorations, The North American Sylva, Thiers' Napoleon, and Macaulay's England. Allibone's magnificent Dictionary of Authors and Books is not yet completed. After the pages have been set and carefully read, they are sent down to the casting-room. Let us go down and see how they fare there.

In the electrotype-room, every thing is as black as the brow of a coal-heaver: in the casting-room, all is as white as the neck of a belle. Take care, sir, or your coat will commit a larceny of our plaster. The form of type is laid on this stone, and nicely oiled: and then a mixture of plaster and water—doesn't it look like a good wife's buckwheat batter?—is poured over it, and gently rolled in. In a short time the plaster sets, and the mould is removed by screws as tenderly as a nurse handles a baby. It is then dried in this hot-tempered oven, and, after the moisture is all evaporated, it is laid in a pan and fastened tightly, as you see, and plunged into this terrible bath

STEREOTYPE CASTING-ROOM.

of a thousand pounds of molten type metal. Phew! you exclaim, what warm work! Yes, sir; but from that fiery sea of lead soon emerges the pan, and its hissing heat is gradually overcome by the water in the trough into which the pan is lowered. Now, caster, break it out. There, Mr. Typograph, is the plate, fixed,—immovable,—stereotyped. The mould is ruined; but the plate is comparatively immortalized. It is rough yet, and, like an uncouth boy, needs polishing.

This next room is the stereotype finishing-room. Here the plates are carefully examined, picked, shaved, trimmed, and boxed, ready for the printer. Take a plate in your hand and examine it: it will bear inspection. You say it is far better than the untrimmed, uneven plates of English founders? We know that, sir; for we have often had to re-finish English plates imported by some publisher who imagined he could save a little by ordering a duplicate set of plates of a popular foreign book. A mistake, sir. Both in type-founding and in stereotyping the Americans have driven the foreigner from the field,—and in the only legitimate way, too: simply by surpassing him.

In this nook on the left, our engraving is done. The drawing is made on the block by the designer, as you see: then patiently and skilfully the engraver cuts and digs out, till the lines and shapes and lights and shades are all revealed in the beautiful picture. Our work in this department gives so much satisfaction that we are seldom without orders.

Now, Mr. Typograph, we shall admit you into our editorial parlour. Walk in, sir. It is not carpeted, and its principal furnishings comprise a desk or two, a few presses, stands and cases, with multitudinous type-surroundings. Here, sir, we edit and print our Specimen Books and our Typographic Advertiser. Don't you see poetical flies buzzing around, and atoms of wit-dust floating in the air, and odours of sentiment stealing out at the key-holes, and grains of common sense sprinkled all over the floor. Will you have a few specimens as curiosities? You say you have already a good assortment in our Advertiser and our Book? Very well, sir: we hope you will treasure them up. You say truly when you remark, that the printing done in this room is seldom, if ever, surpassed in America. We know that; and we intend to stand on the topmost round of the typographical ladder, and to show our fellow-artists what can be done with type such as we manufacture.

We are afraid, Mr. Typograph, that your long excursion over the house has wearied you. Let us get down-stairs again. These, sir, are our warerooms. On these numerous shelves are ranged founts of all the various sorts of types made by us, carefully put up, labelled and classified, and all accessible at a minute's notice. Our customers throughout the country keep actively employed all these porters, packers, clerks, salesmen, and bookkeeper. Many of our customers have never visited us; but we put up their orders with as conscientious fidelity and care as if they were standing before us and watching our every movement. We are happy to see them, and hope none will visit our city without calling in and taking us by the hand. We like to see them face to face, so that we can hang up their portraits in our mental gallery; and, when we afterward receive a letter from them, we can imagine that we are hearing them talk to us rather than reading their writing.

The side-door on which your eye has just rested leads to one of our fire-proofs. Enter it. Here, sir, are safely stored many thousand matrices, as well as moulds, when not in use. As it would require the labour of many weary years to replace them if destroyed, we endeavour to keep them secure from the danger of ruin by fire. The upbuilding of a complete type-foundry is a work of generations.

You will hardly care to look into the basement,—the storehouse of ink and other typographical appliances? Your time is exhausted? Then, sir, we bid you good-day. A safe return to your pleasant family, Mr. Typograph.

4

IMPLEMENTS OR TOOLS OF THE ART

TYPES.

THE types or letters generally used for printing in Europe and America are termed *Roman*, *Italic*, and *Old English*, or *Black Letter*.

ROMAN LETTER.

Roman letter has long been held in the highest estimation, and is the established character of this country, of England, France, Spain, Portugal, and Italy. In Germany, and the kingdoms and states which surround the Baltic, the letters in use are founded on the Gothic character; but, even in those nations, scientific works are printed in their own language with Roman letters.

The Dutch adhere to the black letter in books of devotion and religious treatises; while they make use of the Roman in their curious and learned works.

All printing down to 1465 was in black letter, when characters somewhat improved were introduced at Venice. In 1466, the style known as Roman first appeared, in a volume printed at Rome: this style was brought nearly to its present degree of perfection in Italy as early as 1490.

The Roman letters consist of circles, arcs of circles, and straight lines; and, therefore, on the score of simplicity, precision, and elegance, they certainly deserve to be adopted as the standard for all nations.

A printer, in choosing type, should not only attend to the cut of the letter, but should also observe that its shank is perfectly true, and that it lines or ranges with accuracy, and is of equal height. The quality of the metal of which it is composed, and the finish of the letter, demand particular attention, as the competition for low prices among some of the smaller foundries (which have sprung into existence through the facilities afforded them of multiplying *matrices* by the electrotype process) has induced them to use an inferior metal, and to turn out their letters without due regard to the nicety of finish which is indispensable for proper justification.

It is important that types should have a deep face, their hollows being in proportion to the width of the respective letters, and that the letters should have a deep nick, which should differ from other founts of that body in the same house.

ITALIC LETTER.

For the invention of this letter the world is indebted to Aldus Manutius, by birth a Roman, who erected a printing-office in Venice in 1496, where he introduced Roman types of a neater cut, and gave birth to that striking letter which is known to most nations by the name of Italic; though the Germans and their adherents show themselves as unfair in this respect as they did with the Roman, by calling it Cursiv, *in order to stifle the memory of its original descent, and deprive the Romans of the merit due to the ingenuity of their countryman.*

In the first instance it was termed Venetian, as Manutius was a resident at Venice when he brought it to perfection; but, not long after, it was dedicated to the state of Italy, to prevent any dispute that might arise from other nations claiming a priority.

Italic was originally designed to distinguish such parts of a book as might be considered not strictly to belong to the body of the work, as *Prefaces*, *Introductions*, *Annotations*, &c., all which sub-parts of a work were formerly printed in this character; so that at least two-fifths of a fount was composed of Italic letter.

At present it is used more sparingly, being superseded by the more elegant mode of enclosing extracts within inverted commas, and poetry and annotations in a smaller-sized type. It is often serviceable in distinguishing the head or subject-matter of a chapter from the chapter itself, but is still too often arbitrarily made use of in emphasizing sentences or words.

The frequent use of Italic words among Roman destroys in a great measure the beauty of printing, and often confuses the reader, who, pausing to consider why such words are more strongly noted, loses the context, and must go back to regain the sense of his subject.

BLACK LETTER.

This letter, which was used in the infancy of Printing, descended from the Gothic characters: it is called Gothic by some, and Old English by others; but printers term it Black Letter, on account of its taking a larger compass than either Roman or Italic, the full and spreading strokes thereof appearing more black upon paper.

On the introduction of the Roman character, the use of black letter began to decline, and it was seldom used except in law works, particularly statute law. It was at length expelled from these, and only made its appearance in the heads of law blanks, and as a general display letter.

SAXON CHARACTERS.

The Saxon characters originated probably from the Gothic, but were altered or modified after the Latin ones which the Saxons found in use in England in the fifth century. The first Saxon types were cut by John Daye, under the patronage of Archbishop Parker, about the year 1567. We give the Lord's Prayer in modern Anglo-Saxon types:

Fæꝺeꞃ uꞃe þu þe eaꞃꞇ on heoꝼenum. Si þin nama ᵹehalᵹoꝺ. Tobecume þin ꞃice. Ᵹeƿuꞃðe þin ƿilla on eoꞃþan, ꞅƿa ꞅƿa on heoꝼenum. Uꞃne ꝺæᵹhƿamlican hlaꝼ ᵹẏꝼe uꞅ ꞇo ꝺæᵹ. Anꝺ ꝼoꞃᵹẏꝼ uꞅ uꞃe ᵹẏlꞇaꞅ, ꞅƿa ꞅƿa ƿe ꝼoꞃᵹiꝼað uꞃum ᵹẏlꞇenꝺum. Anꝺ ne ᵹelæꝺꝺe þu uꞅ on coꞅꞇnunᵹe. ac alẏꞅ uꞅ oꝼ ẏꝼele. So ðlice.

NAMES AND SIZES OF TYPE.

THE principal bodies to which printing letters are cast in England and America are the following:—

1. Diamond.	11. English.
2. Pearl.	12. Columbian.
3. Agate.	13. Great Primer.
4. Nonpareil.	14. Paragon.
5. Minion.	15. Double Small Pica.
6. Brevier.	16. Double Pica.
7. Bourgeois.	17. Double English.
8. Long Primer.	18. Double Great Primer.
9. Small Pica.	19. Double Paragon.
10. Pica.	20. Canon.

Besides the foregoing, a smaller size than Diamond, called Brilliant, is now cast in the foundry of Mackellar, Smiths & Jordan, the body of which is just one-half of Minion. Even this is surpassed in smallness by a music type cast in the same foundry, named Excelsior, which is precisely one-half the size of Nonpareil. Another size omitted in the list is Minionette, which is next above Nonpareil, and is largely used for the splendid series of ornamental borders lately originated.

Canon is conceded to have been first produced by a French artisan, and was probably employed in some work relating to the canons of the Church; to which the German title, Missal, alludes.

Two-line Double Pica, Two-line Great Primer, Two-line English, Two-line Pica, and Double Pica, owe their names to the respective bodies of which the depth of two em quadrates answers to one of the double sizes.

Paragon is the only letter that has preserved its name, being called so by all the printing nations. Its appellation shows that it was first cut in France, and at the same time leads us to suppose that the style of letter in that country was at that time but indifferent, and that Paragon happening to turn out a letter of better shape than the rest, it received the name of *perfect pattern*, which the word Paragon implies.

Pica is universally considered as the standard type, and by it furniture is measured, and quotations and labour-saving rules are graduated.

Great Primer, called Tertia in Germany, is one of the major sizes of type which were early used for printing considerable works, and especially the Bible; on which account some persons term it Bible Text. The French name is Gros Romain.

English is called Mittel by the Germans, and St. Augustin by the French and Dutch; the word Mittel (Middle) intimating that the former sizes of letter were seven in number, the centre of which was English, with Prima, Secunda, and Tertia on one side, and Pica, Long Primer, and Brevier on the other. The name St. Augustin was probably given because the writings of that Father were the first works done in that letter.

Pica is called Cicero by the French and Germans. As the preceding size was distinguished by the name of St. Augustine, so this has been honoured with that of Cicero, on account of the Epistles of that writer having been first done in letter of this size. It is doubtful whether the name was given by the French or the Germans.

Small Pica is a grade below Pica, and is now generally employed in octavo volumes, and is, indeed, almost the only size used for printing legal reports and other law books. The French call this letter Philosophie, which, however, is merely a Pica face on a Small Pica body. The Germans term it Brevier.

Long Primer. Upon the supposition that some bodies of letter took their names from works in which they were first employed, we are induced to believe that the Germans gave the name of Corpus to this character on account of their Corpus Juris being first done in this size. The French call this letter Petit Romain.

Bourgeois is a very useful and convenient size of letter. It is frequently used in double-column octavo pages. The name indicates that it originated in France; although type of this body is now called Gaillarde by French printers. Two lines of this letter are equivalent to one line of Great Primer, or four lines of Diamond.

Brevier was first used for printing the Breviaries, or Roman Catholic Church books, and hence its name. The Germans call it Petit, and Jungfer (maiden letter). It is an admirable type, and cannot conveniently be dispensed with in any considerable printing-office.

Minion follows Brevier, and is commonly used for newspapers, and for notes and indexes in book-work. Its name is due probably to its being smaller than any type in use at the period of its invention. It fills a useful place in a printing-office.

Nonpareil came next in order; and its originator, supposing that he had reached the extreme of diminutiveness, gave it this triumphant title. It is extensively used, though mostly on newspapers, and for notes and indexes for duodecimo books and smaller. It is certainly the smallest type that should be allowed in book-work.

Agate probably arose from the necessities of newspaper publishers. As patronage increased, it became desirable to have a type less in size than Nonpareil, for the advertisements, shipping news, markets, &c.; and Agate was made to meet the emergency. It is now extensively used for pocket editions of the Bible and Prayer Books.

Pearl may be said to have been born of ambition. As punch-cutters became more expert, some one possessed of a keen eye and a delicate mechanical finger determined to surpass in smallness the achievements of his predecessors. Hence the origin of this type. This type is also employed in printing miniature volumes.

Diamond followed, as a matter of course; for human ingenuity, when provoked, seems determined to go to the utmost verge of possibility. This type is so minute that a pound of it will contain more than 2300 of the letter i; yet, to produce each letter of an alphabet, a steel punch has to be cut, and a matrix made, in which the types are cast one by one, and, being set up in lines, are rubbed and dressed by the founder for the use of the compositor.

Brilliant. Expert penmen, it is said, have succeeded in writing the Lord's Prayer upon the edge of a sheet of paper. A type-cutter in Berlin, more surprisingly, has formed a type so minute as to be scarcely readable without a good magnifying glass. The type of this paragraph, though not so small as the microscopic letters produced in Prussia, is yet so diminutive that even Diamond is large by comparison. Of the letter i nearly 4000 go to a pound.

GRADATION OF TYPES.

The following specimen shows the proportion which one size of type bears to another in *width;* but it is necessary to observe that it must be taken with certain limitations, because each founder has letter of every size that will either drive out or get in with others of the same body; therefore it is impossible for us to present our readers with a *regular* gradation of the different sizes from *Great Primer* to *Brilliant* inclusive. The limitation of each line is marked by an inverted full-point.

When in the course of human·
When in the course of human· events
When in the course of human· events it b
When in the course of human· events it becomes
When in the course of human· events it becomes nece
When in the course of human· events it becomes necessary
When in the course of human· events it becomes necessary f
When in the course of human· events it becomes necessary for one
When in the course of human· events it becomes necessary for one people to
When in the course of human· events it becomes necessary for one people to dissolve
When in the course of human· events it becomes necessary for one people to dissolve the pol
When in the course of human· events it becomes necessary for one people to dissolve the political bands wh
When in the course of human· events it becomes necessary for one people to dissolve the political bands which have connected th

PROPORTIONS OF TYPES.

It is important for a printer to be acquainted with the exact proportion which one body of letter bears to another in depth. Without this knowledge, he cannot form an accurate judgment as to the size of the type most suitable for a work intended to be confined to a given number of sheets; neither can he form a correct opinion as to the extent of a work, after casting off the copy, unless he possess a rule whereby to guide his calculation as to the quantity of copy which the proposed type may take in. We, therefore, give a scale showing the relative proportions of various letters. Any printer may form a scale by setting up the figures of his various founts, and printing them on a card or dry paper.

Brilliant.	1 2 3 4 5 6 7 8 9 10 11 12 13 14 15 16 17 18 19 20 21 22 23 24 25 26 27 28 29 30 31 32 33 34 35 36 37 38 39 40 41 42 43 44 45 46 47 48 49 50 51 52 53 54 55 56 57 58 59 60 61 62 63 64 65 66 67 68 69 70 71 72 73 74 75 76 77 78 79 80 81 82 83 84 85 86 87 88 89 90 91 92 93 94 95 96 97 98 99 100 101 102 103 104 105
Diamond.	1 2 3 4 5 6 7 8 9 10 11 12 13 14 15 16 17 18 19 20 21 22 23 24 25 26 27 28 29 30 31 32 33 34 35 36 37 38 39 40 41 42 43 44 45 46 47 48 49 50 51 52 53 54 55 56 57 58 59 60 61 62 63 64 65 66 67 68 69 70 71 72 73 74 75 76 77 78 79 80 81 82 83 84 85 86 87 88 89 90 91
Pearl.	1 2 3 4 5 6 7 8 9 10 11 12 13 14 15 16 17 18 19 20 21 22 23 24 25 26 27 28 29 30 31 32 33 34 35 36 37 38 39 40 41 42 43 44 45 46 47 48 49 50 51 52 53 54 55 56 57 58 59 60 61 62 63 64 65 66 67 68 69 70 71 72 73 74 75 76 77 78 79 80
Agate.	1 2 3 4 5 6 7 8 9 10 11 12 13 14 15 16 17 18 19 20 21 22 23 24 25 26 27 28 29 30 31 32 33 34 35 36 37 38 39 40 41 42 43 44 45 46 47 48 49 50 51 52 53 54 55 56 57 58 59 60 61 62 63 64 65 66 67 68 69 70 71 72 73
Nonpareil.	1 2 3 4 5 6 7 8 9 10 11 12 13 14 15 16 17 18 19 20 21 22 23 24 25 26 27 28 29 30 31 32 33 34 35 36 37 38 39 40 41 42 43 44 45 46 47 48 49 50 51 52 53 54 55 56 57 58 59 60 61 62 63 64
Minion.	1 2 3 4 5 6 7 8 9 10 11 12 13 14 15 16 17 18 19 20 21 22 23 24 25 26 27 28 29 30 31 32 33 34 35 36 37 38 39 40 41 42 43 44 45 46 47 48 49 50 51 52 53
Brevier.	1 2 3 4 5 6 7 8 9 10 11 12 13 14 15 16 17 18 19 20 21 22 23 24 25 26 27 28 29 30 31 32 33 34 35 36 37 38 39 40 41 42 43 44 45 46 47 48
Bourgeois.	1 2 3 4 5 6 7 8 9 10 11 12 13 14 15 16 17 18 19 20 21 22 23 24 25 26 27 28 29 30 31 32 33 34 35 36 37 38 39 40 41 42 43 44 45 46
Long Primer.	1 2 3 4 5 6 7 8 9 10 11 12 13 14 15 16 17 18 19 20 21 22 23 24 25 26 27 28 29 30 31 32 33 34 35 36 37 38 39 40
Small Pica.	1 2 3 4 5 6 7 8 9 10 11 12 13 14 15 16 17 18 19 20 21 22 23 24 25 26 27 28 29 30 31 32 33 34 35 36 37
Pica.	1 2 3 4 5 6 7 8 9 10 11 12 13 14 15 16 17 18 19 20 21 22 23 24 25 26 27 28 29 30 31 32
English.	1 2 3 4 5 6 7 8 9 10 11 12 13 14 15 16 17 18 19 20 21 22 23 24 25 26 27 28 29
Great Primer.	1 2 3 4 5 6 7 8 9 10 11 12 13 14 15 16 17 18 19 20 21 22 23

A BILL OF PICA.

THE following is reckoned by the founders a regular fount, perfect in all its sorts:—

A BILL OF 800 LBS. OF PICA.

a	...	8500	,	...	4500	A	...	600	A	...	300
b	...	1600	;	...	800	B	...	400	B	...	200
c	...	3000	:	...	600	C	...	500	C	...	250
d	...	4400	.	...	2000	D	...	500	D	...	250
e	...	12000	-	...	1000	E	...	600	E	...	300
f	...	2500	?	...	200	F	...	400	F	...	200
g	...	1700	!	...	150	G	...	400	G	...	200
h	...	6400	'	...	700	H	...	400	H	...	200
i	...	8000	(	...	300	I	...	800	I	...	400
j	...	400	[	...	150	J	...	300	J	...	150
k	...	800	*	...	100	K	...	300	K	...	150
l	...	4000	†	...	100	L	...	500	L	...	250
m	...	3000	‡	...	100	M	...	400	M	...	200
n	...	8000	§	...	100	N	...	400	N	...	200
o	...	8000	‖	...	100	O	...	400	O	...	200
p	...	1700	¶	...	60	P	...	400	P	...	200
q	...	500				Q	...	180	Q	...	90
r	...	6200	1	...	1300	R	...	400	R	...	200
s	...	8000	2	...	1200	S	...	500	S	...	250
t	...	9000	3	...	1100	T	...	650	T	...	326
u	...	3400	4	...	1000	U	...	300	U	...	150
v	...	1200	5	...	1000	V	...	300	V	...	150
w	...	2000	6	...	1000	W	...	400	W	...	200
x	...	400	7	...	1000	X	...	180	X	...	90
y	...	2000	8	...	1000	Y	...	300	Y	...	150
z	...	200	9	...	1000	Z	...	80	Z	...	40
&	...	200	0	...	1300	Æ	...	40	Æ	...	20
ff	...	400				Œ	...	30	Œ	...	15
fi	...	500	é	...	200						
fl	...	200	à	...	200						
ffl	...	100	â	...	200						
ffi	...	150	ê	...	200						
æ	...	100									
œ	...	60	All other accents, 100 each.								
—	...	150									
——		90									
———		60									

Spaces.

Thick	18000
Middle	12000
Thin	8000
Hair	3000
em Quads	2500
en Quads	5000
Large Quadrates, about 80 lbs.	

Italic, one-tenth of Roman.

Owing to the varying styles of authors and the diverse subjects of books, there will generally be found a number of particular sorts deficient in a fount, whatever the proportions may have been at first. A new fount of letter may run evenly on a work in general literature written in the third person, while a novel filled with dialogues in the first person will rapidly exhaust certain letters, and require sorts to render the fount serviceable to its full general capacity. So with scientific and other books. Even in the case of two authors writing on the same subject, there is no certainty that the fount will run alike. The master-printer, therefore, to keep the entire letter in use, is compelled to order sorts, and his fount is thus constantly growing larger.

A FOUNT OF LETTER.

A COMPLETE fount of letter is comprised under nine heads, in which is contained the following sorts:—

1. *Capitals.*

A B C D E F G H I J K L M N O P Q R S T U V W X
Y Z Æ Œ &.

2. *Small Capitals.*

A B C D E F G H I J K L M N O P Q R S T U
V W X Y Z Æ Œ &.

3. *Lower Case.*

a b c d e f g h i j k l m n o p q r s t u v w x
y z æ œ ff fi ffi fl ffl.

4. *Figures.*

1 2 3 4 5 6 7 8 9 0.

5. *Points, &c.*

, ; : . ? ! - ' ( ) [ ] * † ‡ § ‖ ¶ — ⁓.

6. Four kinds of spaces.
7. Em and en quadrates.
8. Two, three, and four em quadrates.
9. Accents.

These are the ordinary sorts cast to a fount of letter, and are classified by founders as long, short, ascending, descending, kerned, and double letters.

LONG LETTERS fill the whole depth of the body, and are

both ascending and descending, such in the Roman as Q and j, and in the Italic, *f*.

SHORT LETTERS have the face cast on the middle of the body (by founders called shank), as a, c, e, m, n, o, r, s, u, v, w, x, z, all of which will admit of being bearded above and below the face, both in Roman and Italic.

ASCENDING LETTERS are all the Roman and Italic capitals; in the lower case, b, d, f, h, i, k, l, t.

DESCENDING LETTERS are g, p, q, y, in Roman and Italic.

KERNED LETTERS are such as have part of the face hanging over either one or both sides of the body. In Roman, f and j are the only kerned letters; but, in Italic, *d*, *g*, *j*, *l*, *y* are kerned on one side, and *f* on both sides of its face. Most Italic capitals are kerned on one side of the face.

The DOUBLE LETTERS in modern use are ff, fi, ffi, fl, ffl; and these are so cast to prevent the breaking of the beak of the f when used with a tall letter following.

Printers divide a fount of letter into two classes.

1. *The upper case* } *sorts.*
2. *The lower case* }

The upper case sorts are capitals, small capital letters, and references.

The lower case consists of small letters, double letters, figures, points, spaces, quadrates, &c.

CAPITALS.

THE use of capitals in the present day is restricted to proper names of persons, places, &c. There are, however, some authors who deem it essential to mark emphatic words with a capital; in such cases, the author should always send his copy properly prepared in this particular to the printer, or he will become liable to the charge the compositor is allowed to make for loss of time in making any alteration. The method of denoting a capital, or words of capital letters, in manuscript, is by underscoring it with three distinct lines.

SMALL CAPITALS.

SMALL CAPITALS are in general cast to Roman founts only, and are used for the purpose of giving a stronger emphasis to a word than can be conveyed by Italic. They are likewise used for running heads, heads of chapters, &c. The first word of every section or chapter is commonly put in small capitals; but when a two-line initial letter is used, the remainder of the word should be in capitals.

The small capitals C, O, S, V, W, X, Z so closely resemble the same letters in the lower case, that care is required to prevent intermixing.

In manuscript, small capitals are denoted by two lines drawn under the words.

POINTS.

POINTS consist of a comma, semicolon, colon, period or full-point, note of interrogation, and note of admiration.

Points are not of equal antiquity with printing, though, not long after its invention, the necessity of introducing stops or pauses in sentences, for the guidance of the reader, gave birth to the colon and full-point. In process of time, the comma was added, which was then no other than a perpendicular line, proportioned to the body of the letter. These three points were the only ones used till the close of the fifteenth century, when Aldus Manutius, among other improvements in the art of printing, gave a better shape to the comma, and added the semicolon; the comma denoting the shortest pause, the semicolon next, then the colon, and the full-point terminating the sentence. The notes of interrogation and admiration were introduced many years after.

Perhaps there never existed on any subject, among men of learning, a greater difference of opinion than on the true mode of punctuation. Some sprinkle the page with commas almost as promiscuously as if from a pepper-box; some make the pause of a semicolon where the sense will only bear a comma; some contending for what is termed stiff pointing, and others for altogether the reverse.

The want of an established practice is much to be regretted. The loss of time to a compositor, occasioned, often through whim or caprice, in altering points unnecessarily, is a great hardship. Scarcely nine works out of ten are sent properly prepared to the press: either the writing is illegible, the spelling incorrect, or the punctuation defective. Unless the author will take the responsibility of the pointing entirely on himself, it will be to the advantage of the compositor to find not a single point in his copy, unless to terminate a sentence, rather than have his mind confused by commas and semicolons placed indiscriminately, in the hurry of writing, without any regard to propriety.*

The COMMA [,] marks the smallest grammatical division, and commonly represents the shortest pause in reading.

Commas are used to denote extracts or quotations from other works, dialogue matter, or passages or expressions not original, by placing two of them inverted before the first word of the passage quoted, and ending with two apostrophes. A thin space is used to keep the inverted commas free from the matter. The method of running them down the sides to the end of the quotation has been found inconvenient, particularly where a quotation occurs within a quotation, or a speech within a speech; the proper method of distinguishing these is by placing a single inverted comma before such extra quotation, and concluding with a single apostrophe. Where both quotations close together, put three apostrophes, observing after the first to place a thin space.

Inverted commas were first used by Guillemet, a Frenchman, to supersede the use of Italic letter. As an acknowledgment, his countrymen call them after his name. French founders cast them double, thus [« »]. MacKellar, Smiths & Jordan, Philadelphia, furnish them in this way when desired.

A single comma inverted is used as an abbreviation of the word Mac, as in M'Gowen.

The SEMICOLON [;] is used to separate such parts of a sentence as are somewhat less closely connected than those separated by a comma.

The COLON [:] is employed in a sentence between parts less connected than those which are divided by a semicolon, but not so independent as separate, distinct sentences.

The PERIOD OR FULL-POINT [.] serves to indicate the end of a sentence which is independent of any following sentence.

---

* Wilson's *Treatise on English Punctuation* is a full and explicit work on this subject, and should be studied by every printer and author.

When used in abbreviations, it loses its effect as a full stop in the punctuation, unless at the end of a sentence. In some works this point is discarded as a mark of abbreviation, as in Mr Dr &c.

Full-points are sometimes used as leaders in tables of contents, figure-work, &c.; but dotted rules or leaders are much better for this purpose, as they not only supply the place of full-points and quadrates, but save considerable time in the composition.

The sign of INTERROGATION [?] is used to denote a question. It is proper that every interrogation or question should begin with a large letter, whether capitals are used in the matter or not; according to the method observed in the Bible, where interrogatives and responses, and the beginning of sayings, allocutions, &c. are intimated by a capital letter.

The sign of ADMIRATION OR EXCLAMATION [!] denotes surprise, astonishment, rapture, and the like sudden emotions of the mind, whether upon lamenting or rejoicing occasions. This sign is put after the particles Ah! Alas! Oh! &c.; but there are exceptional cases, as, Ah me! Alas the day! &c.

All the points, except the comma and the period, should be preceded by a hair-space; the comma and full-point do not require any space to bear them off.

The em dash [—], though it cannot be denominated a point, is frequently used in peculiar works as a substitute for the comma, or for the colon, and is found particularly serviceable in rhapsodical writing, where interrupted sentences frequently occur.

A dash stands for a sign of repetition in catalogues of goods, where it implies *ditto;* and in catalogues of books, where a dash signifies *ejusdem*, instead of repeating the author's name with the title of every separate treatise of his writing; but no sign of repetition should be at the top of a page, but the name of the author, or of the merchandise, should be set out again at length.

A dash likewise stands for *to*, or *till:* as, chap. xvi. 3-17; that is, from the third to the seventeenth verse inclusive. At other times it serves for an index, to give notice that what follows it is a corollary of what has preceded.

APOSTROPHE.

The apostrophe (') is a comma cast on the upper end of a type, and signifies, where used, the contraction or abbreviation of a word, as is frequently required in poetry to preserve the proper measure of a line. It should not be employed where the primitive word ends with e, as *love*, *change*, &c., but only in cases where the primitive word concludes with a consonant, as, *reign*, *obtain*, &c. It also marks the elision of a vowel at the beginning of words, as, '*scape*, or of a syllable, as, '*prentice*.

The monosyllables *though* and *through* are sometimes shortened to *tho*' and *thro*', but without propriety, as they retain the same sound, and the abbreviation cannot in the slightest degree assist the versification.

Words in the possessive case are generally known by having '*s* for their termination.

All quotations which are denoted at the beginning by inverted commas are closed with apostrophes. There is no space required between the apostrophe and the matter.

HYPHEN, OR DIVISION.

To divide words with propriety is an important part of a compositor's business. It will exercise his judgment, and demands particular attention, as authors must leave the use of the hyphen to the discretion of the printer.

The difficulty that formerly existed as to the proper method of dividing syllables arose from the controversies in which authors were continually engaged on the subject of orthography. Without being able to establish a criterion, each adopted his own particular mode, to the subversion of uniformity and propriety.

A compositor who studies propriety and neatness in his work will not suffer an unnecessary division, even in a narrow measure, if he can avoid it by overrunning two or three lines of matter. In large type and narrow measures, the use of the division may admit of an excuse; but, in that case, care should be taken that hyphens do not follow each other. In small type and wide measures, the hyphen may generally be dispensed with, either by driving out or getting in the word.

without the least infringement on the regularity of the spacing. The compositor who is careful in this respect will find his advantage in the preference given to his work, and in the respect attached to his character as a competent and careful master of his business. Numerous divisions down the side of a page and irregular spacing are the two greatest defects in composition.

It is proper, if possible, to keep the derivative or radical word entire and undivided: as, *occur-rence*, *gentle-man*, *respect-ful*, &c. In other cases, printers generally divide on the vowel, which is an excellent method.

The hyphen is likewise used to connect compound words, which consist frequently of two substantives, as, *bird-cage*, *love-letter*, &c.; likewise what are termed compound adjectives, as, *well-built* house, *handsome-faced* child, &c.

The prepositions *after*, *before*, *over*, &c. are often connected with other words, but do not always make a proper compound: thus, *before-mentioned* is a compound when it precedes a substantive, as, in the *before-mentioned* place; but when it comes after a noun, as, in the place *before mentioned*, it should be two distinct words.*

---

* Wilson, in his *Treatise on English Punctuation*, says, very judiciously,—

The hyphen is employed in words in such a manner as is best calculated to show their origin, composition, or import, and to exhibit the syllables in their neatest form. Agreeably to this rule,—

1. Compound and derivative words are resolved into their primitives; as, *school-master, hand-writing, pen-knife, snuff-box, looking-glass; arch-angel, geo-logy, theo-cracy, ortho-graphy.*

2. Prefixes, affixes, and grammatical terminations, are separated; as, *dis-continue, en-able, trans-port; shear-er, load-ed, print-ing; king-dom, false-hood, differ-ence, command-ment.*

3. One consonant between two vowels is to be joined to the latter syllable, as, *ta-lent, fa-tal; me-lon, le-ver; spi-rit, si-lence; cy-nic, ty-ro; le-ga-cy, mo-no-po-ly.* Except *x* and single consonants when they belong to the former portion of a derivative word; as, *ex-ile, ex-ist, ex-amine; up-on, dis-ease, circum-ambient.*

4. Two or more consonants belong to the latter syllable, when they are capable of beginning a word; as, *ta-ble, sti-fle, lu-cre, o-gle, mau-gre, stro-phe, de-stroy.*

5. But when the consonants cannot begin a word, or when the vowel preceding them is short, the first should be separated; as, *ab-bey, ac-cent, vel-lum, ab-ject, gar-den, laun-dry, pam-phlet; blas-pheme, dis-tress, min-strel.*

It is desirable that compound and derivative words should, at the ends of lines, be divided in such a manner as to indicate their principal parts. Thus, *school-master* is preferable to *schoolmas-ter*, *dis-approve* to *disap-prove*, *resent-ment* to *re-sentment ortho-doxy* to *or-thodoxy;* though, as regards the analysis of words

PARENTHESIS AND BRACKET.

THE use of the PARENTHESIS ( ) is to enclose interpolated words or sentences which serve to strengthen the argument, though the main sentence would read correctly were the enclosed matter taken away.

Parentheses are not now so generally used as formerly: authors place their intercalations between commas, which make them quite as intelligible as though they were enclosed between parentheses, and look much neater in print; but, where parentheses are used, if a point be requisite, it should be placed after the parenthesis, the intercalation not being reckoned any part of that sentence; as, for instance, *My lord (said I), I will tell your lordship, &c.*

Brackets [ ] are seldom made use of now, except to indicate an omitted word inserted to make sense by an editor who scruples to amend some rare composition.

REFERENCES.

REFERENCES are marks and signs employed to direct the reader to the observations which are made in notes at the bottom of a page upon passages of the text to which they are applied.

---

into syllables, the latter mode is unobjectionable. From the narrowness of the printed line, however, in some books, the principle recommended cannot always be adhered to.

The terminations *tion, sion, cial, tial*, and many others, formerly pronounced as two syllables, but now only as one, must not be divided either in spelling or at the end of a line.

A syllable consisting of only one letter, as the *a* in *cre-ation*, should not commence a line. This word would be better divided *crea-tion;* and so all others of a similar kind. But such a syllable, coming immediately after a primitive, is by some printers brought to the beginning; as, *consider-able.*

A line of print must not end with the first syllable of a word when it consists of a single letter; as, *a-bide, e-normous;* nor begin with the last syllable when it is formed of only two letters; as, *nation-al, teach-er, similar-ly.* For regard should be had to the principles of taste and beauty as well as to the laws of syllabication.

Three or more successive lines should not end with a hyphen. A little care on the part of the compositor will, in general, prevent an appearance so offensive to a good eye. Divisions, indeed, except for purposes of spelling and lexicography, should take place as seldom as possible.

References used in works with notes are variously represented, though oftener by letters than other characters. The neatest references, however, when many are required, are either superior letters or superior figures,—thus, ^{1}, ^{2}, ^{3}, or thus, ^{a}, ^{b}, ^{c}. Superior letters are used chiefly in Bibles and other books which have more than one sort of notes, and therefore require various references. When thus used, the letter ^{j} should be omitted, as, from its similarity to the ^{1}, the reader might at times be led into error.

The characters technically known as references by printers are the following:—

Asterisk	*	Double Dagger	‡	Parallel	‖
Dagger	†	Section	§	Paragraph	¶

The Asterisk is the chief reference, and presents itself most readily to the eye. In Roman church-books, the Asterisk divides each verse of a psalm into two parts, and marks where the responses begin, which in the Book of Common Prayer is denoted by a colon placed between the two parts of each verse. Asterisks also denote an omission, or an hiatus by loss of original copy; the number of asterisks being multiplied according to the largeness of the chasm.

The Dagger, originally termed the Obelisk, or Long Cross, is frequently used in Roman Catholic church-books, prayers of exorcism, at the benediction of bread, water, and fruit, and upon other occasions, where the priest is to make the sign of the cross; but the long cross is not used unless for want of square crosses (✠), which are the proper symbols for the purpose. The square cross is used, besides, in the Pope's briefs, and in mandates of archbishops and bishops, who put it immediately before the signature of their names; but it is not reckoned among references.

The Paragraph was formerly prefixed to such matter as authors designed to distinguish from the mean contents of their works, and to give the reader an item of some particular subject. At present, paragraphs are used chiefly in Bibles, to show the parts into which a chapter is divided. In Common Prayer Books, paragraphs are put before the matter that directs the order of the Service, and which is called the Rubric because those lines were formerly printed in red.

ACCENTED LETTERS.

THOSE which are called accented by printers are the five vowels, marked thus:—

Acute	á é í ó ú
Grave	à è ì ò ù
Circumflex....................................	â ê î ô û
Diæresis	ä ë ï ö ü
Long..	ā ē ī ō ū
Short	ă ĕ ĭ ŏ ŭ

We may include the French ç, the Spanish ñ, the Portuguese õ or ẽ, and the Welsh ŵ and ŷ.

NUMERAL LETTERS.

THE Greeks at first employed the letters of the entire alphabet to express the first twenty-four numbers; but the system was cumbrous, and they adopted the happy expedient of dividing their alphabet into three portions, using the first to symbolize the 9 digits, the second the 9 tens, and the third the 9 hundreds; and, as their alphabet contained only twenty-four letters, they invented three additional symbols. Their list of symbols then stood as follows:—

Units.		Tens.		Hundreds.	
α represents..........	1	ι represents..........	10	ρ represents	100
β	2	κ	20	σ	200
γ	3	λ	30	τ	300
δ	4	μ	40	υ	400
ε	5	ν	50	φ	500
ϛ (introduced).........	6	ξ	60	χ	600
ζ	7	ο	70	ψ	700
η	8	π	80	ω	800
θ or ϑ	9	ϟ or ϙ (introduced)	90	ϡ, ϡ, ϡ (introd'd)	900

By these symbols, only numbers under 1000 could be expressed; but, by putting a mark called iota under any symbol, its value was increased a thousand-fold: thus, ᾳ = 1000, ϛ = 20,000; or, by subscribing the letter M, the value of a symbol was raised ten thousand-fold. For these two marks, single and double dots were afterward substituted. This improvement enabled them to express with facility all numbers as high as 9,990,000,—a range amply sufficient for all ordinary purposes.

It has been supposed that the Romans used M to denote 1000, because it is the first letter of Mille, which is Latin for 1000; and C to denote 100, it being the first letter of Centum, the Latin term for 100. Some also suppose that D, being formed by dividing the old M in the middle, was therefore appointed to stand for 500,—that is, half as much as the M stood for when it was whole; and that L being half a C, was, for the same reason, used to denominate 50. But the most natural account of the matter appears to be this:—

The Romans probably put down a single stroke, I, for one, as is still the practice of those who score on a slate, or with chalk; this stroke they doubled, trebled, and quadrupled, to express two, three, and four: thus, II, III, IIII. So far they could easily number the minums or strokes with a glance of the eye; but they found that if more were added it would be necessary to number the strokes one by one: for this reason, when they came to five, they expressed it by joining two strokes together in an acute angle, thus, V, which will appear the more probable if it be considered that the progression of the Roman numbers is from five to five,—that is, from the fingers of one hand to the fingers of the other.

After they had made this acute angle, V, for five, they then added single strokes to the number of four, thus, VI, VII, VIII, VIIII, and then, as the minums could not be further multiplied without confusion, they doubled their acute angle by prolonging the two lines beyond their intersection, thus, X, to denote two fives, or ten. After they had doubled, trebled, and quadrupled this double acute angle, thus, XX, XXX, XXXX, they then, for the same reason which induced them to make a single angle first, and then to double it, joined two single strokes in another form, and, instead of an acute angle, made a right angle, L, to denote fifty. When this was doubled, they then doubled the right angle, thus, ⊏, to denote one hundred, and, having numbered this double right angle four times, thus, ⊏⊏, ⊏⊏⊏, ⊏⊏⊏⊏, when they came to the fifth number, as before, they reverted it, and put a single stroke before it, thus, I⊐, to denote five hundred; and, when this five hundred was doubled, then they also doubled their double right angle, setting two double right angles opposite to each other, with a single stroke between them, thus, ⊏I⊐, to denote one thousand; when this note for one thousand had been repeated four times, they then put down I⊐⊐ for five thousand, ⊏⊏I⊐⊐ for ten thousand, and I⊐⊐⊐ for fifty thousand.

The corners of the angles being cut off by transcribers for despatch, these figures were gradually brought into what are now called numerical letters. When the corners of ⊏I⊐ were made round, it stood thus, CIƆ, which is so near the Gothic 𝔪 that it soon deviated into that character; so that I⊐ having the corners made round, stood thus, IƆ, and then easily deviated into D. ⊏ also became a plain C by the same means: the single rectangle, which denoted fifty, was, without any alteration, a capital L; the double acute angle was an X; the single acute angle, a V consonant; and a plain single stroke, the letter I. And thus these seven letters, M, D, C, L, X, V, I became numerals. As a further proof of this assertion, let it be considered that CIƆ is still used for one thousand, and IƆ for five hundred, instead of M and D; and this mark, 𝔪, is sometimes used to denote one thousand, which may easily be derived from this figure, ⊏I⊐, but cannot be deviations from, or corruptions of, the Roman letter M. The Romans also expressed any number of thousands by a line drawn over any numeral less than one thousand: thus, $\overline{V}$ denotes five thousand, $\overline{LX}$ sixty thousand; so, likewise, $\overline{M}$ is one million, $\overline{MM}$ two millions, &c.

Upon the discovery of printing, and before capitals were invented, small letters served for numerals; not only when Gothic characters were in vogue, but when Roman had become the prevailing character. Thus, in early times, 𝔦 𝔳 𝔵 𝔩 𝔠 𝔡 𝔪 were, and in Roman type are still, of the same signification as capitals when used as numerals. Though the capital J is not a numeral letter, yet the lower-case 𝔧 is as often and as significantly used as the vowel 𝔦, especially where the former is employed as a closing letter, in 𝔦𝔧 𝔦𝔦𝔧 𝔳𝔧 𝔳𝔦𝔧 𝔳𝔦𝔦𝔧 𝔡𝔠𝔦𝔧, &c. In Roman lower-case numerals, the j is not regarded, but the i stands for figure 1 wherever it is used numerically.

During the existence of the French Republic, books were dated in France from the first year of the Republic: thus, Ann. XII. (1803), or twelve years from 1792.

ARITHMETICAL FIGURES.

THE arithmetical or Arabic numerals are 0, 1, 2, 3, 4, 5, 6, 7, 8, 9. Properly they should be styled Hindu or Indian numerals; for the Arabs borrowed them, along with the decimal system of notation, from the Hindus. They were probably first introduced from the East into Italy about 1202; yet they did not come into general use before the invention of printing. Accounts were kept in Roman numerals up to the sixteenth century.

OLD STYLE FIGURES.

THOUGH uniform in height and appearance, we do not deem the modern figures an improvement on the variously-lining figures formerly in vogue, and now happily coming again into use. The latter can be caught by the eye with greater ease and certainty, just as lower-case letter can be read with more facility than continuous lines of capitals. In the new style the 3 and 8 may easily be mistaken for each other, and so with the 6, 9, and 0; but in the non-lining figures such errors are quite unlikely to happen, as some of them occupy the centre of the body only, and others are ascending or descending characters. The example here given will show the justice of our remarks:

1 2 3 4 5 6 7 8 9 0
1 2 3 4 5 6 7 8 9 0

SCRATCHED OR CANCELLED FIGURES

1 2 3 4 5

ARE used in arithmetical matter when the divided and dividing figures require to be crossed over in an operation.

FRACTIONS.

COMMON FRACTIONS, or broken numbers in arithmetic, are cast to all sizes of type solid. A great improvement has recently been introduced by casting the numerator and denominator separately, on bodies of half size, with the line on the under figure, so that odd fractions of any amount may be easily formed.

SIGNS.

COMMERCIAL SIGNS.

℔ ... *Per*, each.
@ ... At or to.
% ... Per centum.
℅ ... account.
¢ ... cent.
$ ... Dollar or dollars.
£ ... *Libra*, *libræ*, pound or pounds sterling.
/ ... *Solidus*, *solidi*, shilling or shillings.

MATHEMATICAL, ALGEBRAICAL, AND GEOMETRICAL.

+ *plus*, or *more*, is the sign of real existence of the quantity it stands before, and is called an affirmative or positive sign. It is also the mark of addition: thus, $a + b$, or $6 + 9$, implies that a is to be added to b, or 6 added to 9.

− *minus*, or *less*, before a single quantity, is the sign of negation, or negative existence, showing the quantity to which it is prefixed to be less than nothing. But between quantities it is the sign of subtraction: thus, $a - b$, or $8 - 4$, implies b subtracted from a, or 8 after 4 has been subtracted.

= *equal*. The sign of equality, though Des Cartes and some others use this mark, ∞: thus, $a = b$ signifies that a is equal to b. Wolfius and some others use the mark = to denote the identity of ratios.

× *into* or *with*. The sign of multiplication, showing that the quantities on each side the same are to be multiplied by one another: as, $a \times b$ is to be read, a multiplied into b; 4×8, the product of 4 multiplied into 8. Wolfius and others make the sign of multiplication a dot between the two factors: thus, $7 . 4$ signifies the product of 7 and 4. In algebra the sign is commonly omitted, and the two quantities put together: thus, bd expresses the product of b and d. When one or both of the factors are compounded of several letters, they are distinguished by a line drawn over them: thus, the factum of $a + b - c$ into d is written, $d \times \overline{a + b - c}$. Leibnitz, Wolfius, and others distinguished the compound factors by including them in a parenthesis: thus, $(a + b - c)\, d$.

÷ *by*. The sign of division: thus, $a \div b$ denotes the quantity a to be divided by b. Wolfius makes the sign of division two dots; $12 : 4$ denotes the quotient of 12 divided by $4 = 3$. If either the divisor, or dividend, or both, be composed of several letters, for example, $a \,..\, b \div c$, instead of writing the quotient like a fraction.

7 or ⊏ are signs of majority: thus, a 7 b expresses that a is greater than b.

∠ or ⊐ are signs of minority,—when we would denote that a is less than b.

∽ is the character of similitude used by Wolfius, Leibnitz, and others. It is used in other authors for the difference between two quantities when it is unknown which is the greater of the two.

: : *so is.* The mark of geometrical proportion disjunct, and is usually placed between two pair of equal ratios: as, 3 : 6 : : 4 : 8 shows that 3 is to 6 as 4 is to 8.

: or ∴ is an arithmetical equal proportion: as, 7 . 3 : 13 . 9; *i.e.* 7 is more than 3, as 13 is more than 9.

□ quadrate, or regular quadrangle,—viz. □ AB = □ BC; *i.e.* the quadrangle upon the line AB is equal to the quadrangle upon the line BC.

△ triangle: as, △ ABC = △ ADC.

∠ an angle: as, ∠ ABC = ∠ ADC.

⊥ perpendicular: as, AB ⊥ BC.

▭ rectangled parallelogram, or the product of two lines.

‖ the character of parallelism.

ǂ want of parallelism.

≚ equiangular, or similar.

≟ equilateral.

▱ rhomboid.

⌒ concentric.

○ circle.

∟ right angle.

° denotes a degree: thus, 45° implies 45 degrees.

′ a minute: thus, 50′ is 50 minutes; ″, ‴, ⁗, denote seconds, thirds, and fourths; and the same characters are used where the progressions are by tens, as it is here by sixties.

∺ the mark of geometrical proportion continued, implies the ratio to be still carried on without interruption: as, 2, 4, 8, 16, 32, 64 ∺ are in the same uninterrupted proportion.

√ *irrationality.* The character of a surd root, and shows, according to the index of the power that is set over it or after it, that the square, cube, or other root is extracted, or to be ex-

METAL RULES OR DASHES.

Metal Rules or dashes, like quadrates, are commonly cast from an em to three ems in length. When cast to line and join accurately, they may serve instead of brass rule.

BRACES.

Braces are chiefly used in tables of accounts, botanical and geological tables, and similar matter. They stand *before* and keep together items of similar import, or subdivisions of the preceding article. They sometimes stand *after* and keep together such articles as make more than one line, and have either pecuniary, mercantile, or other posts after them, which are justified to answer to the middle of the brace. Braces are sometimes used horizontally in the margin, to cut off a chronological or other series from the proper notes or marginal references of the work.

Braces are generally cast to two, three, and four ems of each fount. *Middles* and *ends* are also cast, which can be filled out with dashes to any length required for the brace.

Middles and ends are convenient in genealogical works, where they are used the flat way, and where the directing point is not always in the middle, but has its place under the name of the parent, whose offspring stands between corner and corner of the brace inside, in order of primogeniture.

Brass braces of any length, for music and jobbing purposes, are furnished by type-founders.

SPACES.

Spaces are short blank types, and are used to separate one word from another. To enable the compositor to space even and to justify with nicety, they are cast to various thicknesses, —viz. five to an em, or five thin spaces; four to an em, or four middlé spaces; three to an em, or three thick spaces; and two to an em, or two en quadrates, which may with propriety be reckoned among the number of spaces. Besides these, there is what is called the hair-space, cast remarkably thin, and found particularly useful in justifying lines and assisting uniformity in spacing.

TWO-LINE LETTERS

ARE equal in depth to two lines of the type in which they are to be used, and of proportionate width. They form the almost only proper type for title-pages, and are used at the beginning of chapters and newspaper advertisements.

QUADRATES.

AN em quadrate is a short blank type, in thickness equal to the square of the letter of the fount to which it belongs; an en quadrate is half that size. In casting em and en quadrates, the utmost exactness is necessary; they also require particular care in dressing, as the most trifling variation will instantly be discovered when they are ranged in figure-work, and, unless true in their justification, the arrangement will be so irregular that all the pains and ingenuity of a compositor cannot rectify it. The same observation will hold good with respect to figures.

The first line of a paragraph is usually indented an em quadrate; but some printers prefer using an em and en, two, or even three ems for wide measures. An em quadrate is the proper space after a full-point when it terminates a sentence in a paragraph.

En quadrates are generally used after the semicolon, colon, &c., and sometimes after an overhanging letter. They are useful in spacing.

The inconvenience arising from founts of the same body not agreeing in depth is great, where the quadrates, through necessity, are sometimes mixed. The founts cast by L. Johnson & Company are not liable to this charge, as their moulds for all regular type of a specific size harmonize perfectly, and the quadrates and spaces work together.

QUOTATIONS.

QUOTATIONS are cast to two sizes, and are called broad and narrow. They vary in size according to the standard of the foundry where they are cast. They are being superseded, however, by

LABOUR-SAVING QUOTATION FURNITURE.

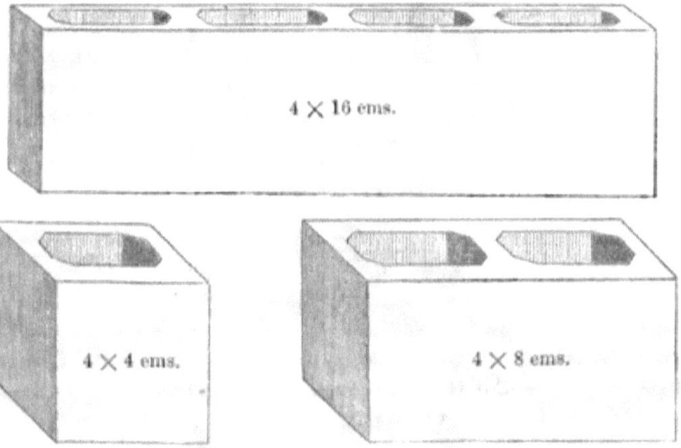

THIS is cast with great accuracy to Pica, of assorted widths and lengths; and, as its name imports, it serves not only for quotations in general job-work, but also for furniture: not being liable to warp or shrink, it forms a highly economical substitute for wooden reglet.

ANSWER many of the purposes of quotations, but are principally useful as frames or miniature chases for circular or oval jobs. They are cast of various sizes, graduated to Pica ems.

CIRCULAR QUADRATES.

These are made of various sizes, so as to form circles or parts of circles from one to twenty-four inches in diameter. Each piece is exactly one-eighth of a full circle, and, when combined with similar pieces, will form quarter, half, three-quarter, and full circles. By reversing the combination of some of the pieces, serpentine and eccentric curves may be made of any length or depth.

There are two kinds: *inner quadrates*, with convex surface, and *outer quadrates*, with concave surface. The curved line is produced by placing the convex and concave surfaces parallel to each other, so that when locked up firmly they hold the type inserted between them. The other sides of the quadrates are flat and right-angled, to allow a close introduction of type, and an easy justification with common quadrates.

DIRECTIONS FOR COMPOSING CIRCULAR QUADRATES FOR THE ORDINARY SIZE OF ONE-FOURTH CIRCLE.

Select two outer quadrates (each marked with the same number) of the length required. Join the smaller ends, and justify the extremities carefully with ordinary quadrates. Set the line of type in the hollow of the curve, but without justification. Then insert two inner quadrates (of the same number) of smaller size than the outer quadrates. The size of the inner quadrates will depend upon the size of the type. A line of Canon will require smaller inner quadrates than will be needed for a line of Pica, and *vice versâ.* As the one increases, the other diminishes. An ordinary clock-dial will afford a good illustration. The space between the numerals X and I is one-fourth of a circle. The curved line described around the foot of these numerals is much less than the curve at the top. If the size of the numerals from X to I is decreased, the inner curve will be greater; if it is increased, it will be less. This will explain why the inner quadrate should be of less size than the outer, and why it should diminish as the size of the type increases. The curve of the inner quadrate should be perfectly parallel with the curve of the outer quadrate. When they are parallel, they bind the type between firmly in every part. Then justify the line of type.

As the sizes of type vary with different foundries, it will often be found that the inner quadrate of the nearest suitable size will not meet the type in every part. This difficulty may be obviated by introducing slips of card *of the same length as the line of type.* Thus increase the distance between the quadrates until the curved surfaces are perfectly parallel with each other. The line of type cannot be justified unless they are parallel. When the inner and outer quadrates are thus adapted to each other, they will not only bind the type firmly, but will also present a perfectly flat and square surface on the other side. Unless they are parallel on the inner sides, and flat and square on the outer sides, the justification is not good; and the remedy must be found in changing the size of the inner circle, or in increasing the distance between the curved lines by the use of larger type or by paper or card board. When thus composed, the type will be perfectly tight and secure, and the curved white line strictly accurate.

As these quadrates are perfect segments of a large circle, they cannot be increased or diminished without destroying the truth of the curve. If the thin ends are pieced out with common quadrates, good justification will be rendered impossible; if they are shortened by cutting off, they are ruined. Bits of lead or short pieces of card between the curved surfaces are also wrong: they destroy that exact parallelism which is necessary for the security of the type. Very accurate justification of the outer extremities of the quadrates is also indispensable. If the curved surfaces are kept parallel, and the flat surfaces kept square, no difficulty will be found in using them, and they will prove a valuable aid in ornamental printing.

LEADS.

LEADS form a very important part of a printer's stock in trade, since it is scarcely possible to set up a single page in which they may not be usefully employed; but their chief use is for opening the lines to a regular distance from each other. They are usually cast by letter-founders in a long mould, and then cut to the required lengths. The bodies are regulated by Pica standard, and they are usually cast four, six, or eight to Pica, but are occasionally varied from one down to fourteen to Pica. The lengths also vary,—twenty ems Pica being about the average; though they are cut to almost every length, in order that, by being combined, they may suit every measure.

FLOWERS AND BORDERS.

THE flowers and borders designed and cast at the present time far surpass any made by founders of earlier days. Their richness and variety enable printers to execute delicate and elaborate work rivalling plate engraving, and afford a wide scope for the display of taste and artistic skill. The combination borders are especially valuable from being cast on uniform bodies, thus rendering them susceptible of a vast number of changes.

The ancient practice of ornamenting pages with head and tail pieces of flowers seems to be coming in vogue again, particularly in works printed with old-style type, such as was used a century ago.

BRASS RULES.

RULES are required mainly for table-work, and for pages which contain two or more columns. They are also useful in titles and jobs. Brass rule should never be more than type height, unless for perforating purposes, to divide railroad checks, &c. A shade lower would be often better, as the pressman would be enabled to bring off their impression more clearly. In table-work, the rule and figures should be separated by a lead, and all the rules should fit closely and accurately.

The lately-invented

BRASS LABOUR-SAVING RULE

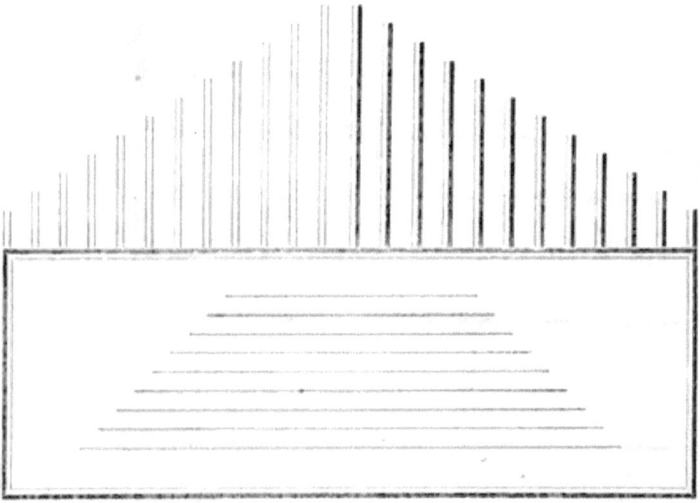

is of immense economical advantage to the printer. Being cut to a graduated scale, from one em to fifty ems Pica in length, advancing in the shorter pieces by ens and in the longer by ems, all wastage in cutting is avoided by the printer, as rules of any length can be formed by employing two or more pieces. They are put up in regular founts, of various styles, with sufficient mitred pieces for outside bordering.

Metal space-rules, cast by type-founders, are commonly used for cross-rules in table-work; but the shorter pieces of labour-saving rule will answer as well.

On the next page we insert a plan of a case for labour-saving rules, with boxes suited for the various sizes, in which the rule should be kept when not in use.

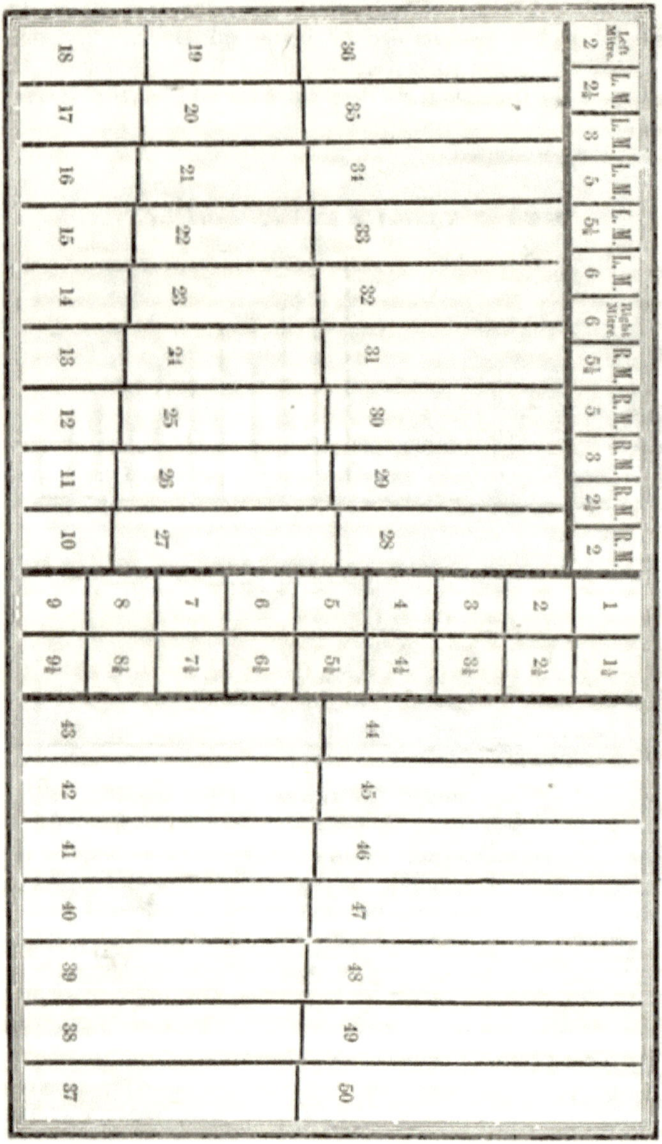

LABOUR-SAVING RULE CASE.

RUNIC ALPHABETS.

Runes were the earliest alphabets in use among the Teutonic and Gothic nations of Northern Europe. The exact period of their origin is not known. The name is derived from the Teutonic *rûn*, a mystery; whence *runa*, a whisper, and *helrûn*, divination; and the original use of these characters seems to have been for purposes of secrecy and divination. Scandinavian and Anglo-Saxon tradition agree in ascribing the invention of runic writing to Odin or Wodin. The countries in which traces of the use of runes exist include Denmark, Norway, Sweden, Iceland, Germany, Britain, France, and Spain; and they are found engraved on rocks, crosses, monumental stones, coins, medals, rings, brooches, and the hilts and blades of swords. Runic letters were also often cut on smooth sticks called *rûn-stafas*, or mysterious staves, and used for purposes of divination. But there is no reason to believe that they were at any time in the familiar use in which we find the characters of a written language in modern times, nor have we any traces of their being used in books or on parchment. We have an explanation of the runic alphabet in various MSS. of the early middle ages, prior to the time when runes had altogether ceased to be understood.

The systems of runes in use among the different branches of the Teutonic stock were not identical, though they have a strong general family likeness, showing their community of origin. The letters are arranged in an order altogether distinct from that of any other alphabetical system, and have a purely Teutonic nomenclature. Each letter is, as in the Hebrew-Phœnician, derived from the name of some well-known familiar object, with whose initial letter it corresponds. Runes, being associated in the popular belief with augury and divination, were to a considerable extent discouraged by the early Christian priests and missionaries, whose efforts were directed to the supplanting of them by Greek and Roman characters. But it was not easy suddenly to put a stop to their use, and we find runes continuing to be employed in early Christian inscriptions. This was to a remarkable extent the case in the Anglo-Saxon kingdoms of Northumbria, Mercia, and East Anglia, where we have traces of runic writing of dates varying from the middle of the seventh to the middle of the tenth century. Runes are said to have been laid aside in Sweden by the year 1001, and in Spain they were officially condemned by the Council of Toledo in 1115.

The different systems of runes, all accordant up to a certain point, have been classed as the Anglo-Saxon, the German, and the Norse, each containing different subordinate varieties. The Norse alphabet is generally considered the oldest, and the parent of the rest. It has sixteen letters corresponding to our *f, u, th, o, r, k, h, n, i, a, s, t, b, l, m, y*, but has no equivalent for various sounds which existed in the language, in consequence of which the sound of *k* was used for *g*, *d* for *t*, *b* for *p*, and *u* and *y* for *v*: *o* was expressed by *au*, and *e* by *ai*, *i*, or *ia*; and the same letter otherwise was made to serve for more than one sound. Other expedients came, in the course of time, to be employed to obviate the deficiency of the system,—as the addition of dots, and the adoption of new characters. But the runic system received a fuller development among the Germans and Anglo-Saxons, particularly the latter, whose alphabet was extended to no fewer than forty characters, in which seem to have been embraced, more nearly than in any modern alphabets, the actual sounds of a language. The table on the following page exhibits the best known forms of the Anglo-Saxon, German, and Norse runic alphabets, with the names and the power of the several letters.

The Anglo-Saxon runes, as here given, are derived from a variety of MS. authorities, the most complete containing forty characters, while some only extend as far as the twenty-fifth or twenty-eighth letter. Neither the name nor the power of some of the later letters is thoroughly known, and they are without any equivalents in the Norse runic system. The German runes are given from a MS. in the conventual library of St. Gall, in Switzerland. Though the various runic alphabets are not alike copious, the same order of succession among the letters is preserved, excepting that, in the Norse alphabet, *laugr* precedes *madr*, although we have placed them otherwise, with the view of exhibiting the correspondence of the three systems. The number of characters in the Anglo-Saxon alphabet is a multiple of the sacred number eight; and we have the evidence both of a Swedish bracteate containing twenty-four characters, and of the above mentioned St. Gall MS., that there was a recognised division of the alphabet into classes of eight letters,—a classification which forms the basis

Anglo-Saxon.			German.		Norse.	
ᚠ	feoh	f	ᚠ	feh	ᚠ	fé
ᚢ ᚣ	ur	u (short)	ᚢ	uur	ᚢ	ur
ᚦ	thorn	th	ᚦ	dorn	ᚦ	thurs
ᚩ ᚨ	os	o (short)	ᚩ	oos	ᚬ	os
ᚱ	rad	r	ᚱ	rat	ᚱ	ridr
ᚳ	cæn	k	ᚳ	cen	ᚴ	kaun
ᚷ	gyfu	g	ᚷ	gebo		
ᚹ	wen	w	ᚹ	huun		
ᚻ	hægel	h	ᚻ	hagal	ᚼ	hagl
ᚾ	nyd	n	ᚾ	nod	ᚾ	naud
ᛁ	is	i (short)	ᛁ	iis	ᛁ	is
ᛄ	gear	y (cons.)	ᛄ	ger	ᛅ	ar
ᛇ	eoh	e (long)	ᛇ	ih		
ᛈ ᛕ	peorth	p	ᛈ	perd		
ᛉ	eolhx	x	ᛉ	elix		
ᛋ	sigel	s	ᛋ	sigi	ᛋ	sol
ᛏ	tir	t	ᛏ	ti	ᛏ	tyr
ᛒ	beorc	b	ᛒ	borg	ᛒ	biarkan
ᛖ	eh	e (short)	ᛖ	eh		
ᛗ	man	m	ᛗ ᛗ	man	ᛘ	madr

Anglo-Saxon.			German.		Norse.	
ᛚ	lagu	l	ᛚ	lago	ᛚ	laugr
ᛝ ᛝ	ing	ng	ᛝ	inc		
ᛞ	dæg	d	ᛞ	tag		
ᛟ ᛟ	œthel	o (long)	ᛟ	odil		
ᚪ ᚪ	ac	a (long)	ᚪ	ac	ᛦ	yr
ᚫ	æsc	a (short)	ᚫ	asc		
ᚣ ᚣ	yr	y	ᚣ	yur		
ᛠ	ear	au	ᛠ	der		
ᛡ	ior	io				
ᛢ ᛢ	queorn	q				
ᛣ ᛤ	calc					
ᛥ	stan	st				
ᚸ ᚸ	gar	dzh				
Z		z				
[illegible]						
[illegible]						
[illegible]	vult	v				
[illegible]						
[illegible]						
[illegible]						

of a system of secret runes mentioned in that MS. Of these secret runes, there are several varieties specified: in particular, 1. *Iis-runa* and *Lago-runa* (of which specimens exist in Scandinavia), consisting of groups of repetitions of the character *iis* or *lago*, some shorter and some longer, the number of shorter characters in each group denoting the class to which the letter intended to be indicated belonged; the number of longer ones, its position in the class. 2. *Hahal-runa*, where the letters are indicated by characters with branching stems, the branches to the left denoting the class, and those to the right the position in that class. There is an inscription in secret runes of this description at Hackness, in Yorkshire. 3. *Stof-runa*, in which the class is indicated by points placed above, and the position in the class by points below, or the reverse.

The best-known inscriptions in the Anglo-Saxon character are those on two grave-stones at Hartlepool, in Northumberland, on a cross at Bewcastle, in Cumberland, and on another cross at Ruthwell, in Dumfriesshire. The inscription on the west side of Bewcastle cross, which we give as a specimen of Anglo-Saxon runes, is a memorial of Alcfrid, son of Oswiu, who was associated with his father in the government of the kingdom of Northumbria, in the seventh century.

RUNES.

It has been thus deciphered into the Anglo-Saxon dialect of the period:—

+THIS SIGBECUN
SETTÆ HWÆTRED
EM GÆRFÆ BOLDU
ÆFTÆR BARÆ
YMB CYNING ALCFRIDÆ
GICEGÆD HEOSUM SAWLUM.

Or, in modern English:—

This memorial
Hwætred set
and carved this monument
after the prince
after the king Alcfrid,
pray for their souls.

The inscription on the Ruthwell cross, after being long a puzzle to antiquaries, was first deciphered in 1838 by Mr. John

M. Kemble, an eminent Anglo-Saxon scholar. It is written alternately down one side of the stone and up another, and contains a portion of a poem on the subject of the Crucifixion. Mr. Kemble's interpretation received a very satisfactory confirmation by the discovery of a more complete copy of the same poem in a MS. volume of Anglo-Saxon homilies at Vercelli.

Mr. D. M. Haigh, whose researches have added much to our knowledge of Anglo-Saxon runes, has endeavoured to set up for them a claim of priority over the Norse characters. Instead of considering the additional Anglo-Saxon letters as a development of the Norse system, he looks on the Norse alphabet of sixteen letters as an abridgment of an earlier system, and finds occasional traces of the existence of the discarded characters in the earliest Norse inscriptions, and in the Scandinavian *Iis-runa* and *Hahal-runa*, where the letters are classified in accordance with the Anglo-Saxon groups of eight.

The Scandinavian kingdoms contain numerous runic monuments, some of them written *boustrophedon*, or with the lines beginning alternately from the right and left; and there are many interesting inscriptions on Swedish gold bracteates, generally having reference to some design which they accompany. The Celtic races, from their connection with the Scandinavians, became acquainted with their alphabet, and made use of it in writing their own language; and hence we have, in the Western Islands of Scotland and in the Isle of Man, runic inscriptions, not in the Anglo-Saxon, but in the Norse character, with, however, a few peculiarities of their own. Some of the most perfect runic inscriptions are in Man; others of a similar description exist at Holy Island, in Lamlash Bay, Arran; and there is an inscription in the same character on a remarkable brooch dug up at Hunterston, in Ayrshire. Dr. D. Wilson considers that the Celtic population of Scotland were as familiar with Norse as the Northumbrians with Saxon runes.

We sometimes find the Norse runes used to denote numerals, in which case the sixteen characters stand for the numbers from 1 to 16; *ar* combined with *laugr* stands for 17, double *madr* for 18, and double *tyr* for 19. Two or more letters are used to express higher numbers, as *ur ur*, 20; *thurs thurs os*, 34.*

* Chambers's Encyclopædia, published by J. B. Lippincott & Co., Philadelphia.

ANGLO-SAXON ALPHABET.

THE Anglo-Saxon alphabet, and the forms and sounds of the letters, are shown in the following table:—

Form.	Sound.	Form.	Sound.
A ... a	*a* as in *bar*	N ... n	*n*
B ... b	*b*	O ... o	*o*
C ... c	*c* as in *choice*	P ... p	*p*
D ... ð	*d*	R ... ր	*r*
E ... e	*e* as in *feint*	S ... ſ	*s*
F ... ꝼ	*f*	T ... ꞇ	*t*
G ... ᵹ	*g* as in *gem*	Ð Þ ... ð þ	*th*
H Ƕ ... h	*h*	U ... u	*u*
I ... ı	*i*	W ... ƿ	*w*
K ... k	*k*	X ... x	*x*
L ... l	*l*	Y ... ẏ	*y*
M ... m	*m*	Z ... z	*z*

Two useful Anglo-Saxon letters have disappeared from modern English,—namely, Þ or þ th (*th*in), and Ð or ð th (*th*ine).

The Anglo-Saxon letters which vary from those now used were doubtless mere corruptions of the Roman forms,—viz. the capitals A, C, E, G, H, M, S, and W, and the small letters d, f, g, i, r, s, t, and w. Several notes of abbreviation were used by the Saxons, as ꝥ *that*, ⁊ *and*, &c. The notes of abbreviation are not original members of the alphabet: they were the result of later reflection, and were introduced probably for despatch.

About the year 1567, John Daye, who was patronized by Archbishop Parker, cut the first Saxon types which were used in England. In this year, *Asserius Menevensis* was published by the direction of the archbishop in these characters; in the same year, Archbishop Ælfric's *Paschal Homily;* and in 1571, the *Saxon Gospels.*

On the two following pages will be found a plan of cases for Saxon types.

SAXON UPPER CASE.

*	†	‡	§	‖	¶	☞							
									2 em ⏟	3 em ⏞	⏜	⏞	⏜
		⁊ *					-	—	2 em ——	3 em ———			
A	B	C	D	E	F	G	æ						
H	I	K	L	M	N	O	þ						
P	Q	R	S	T	Þ	Ƿ							
X	Y	Z	Ð	U	]	)						hair space.	

SAXON LOWER CASE.

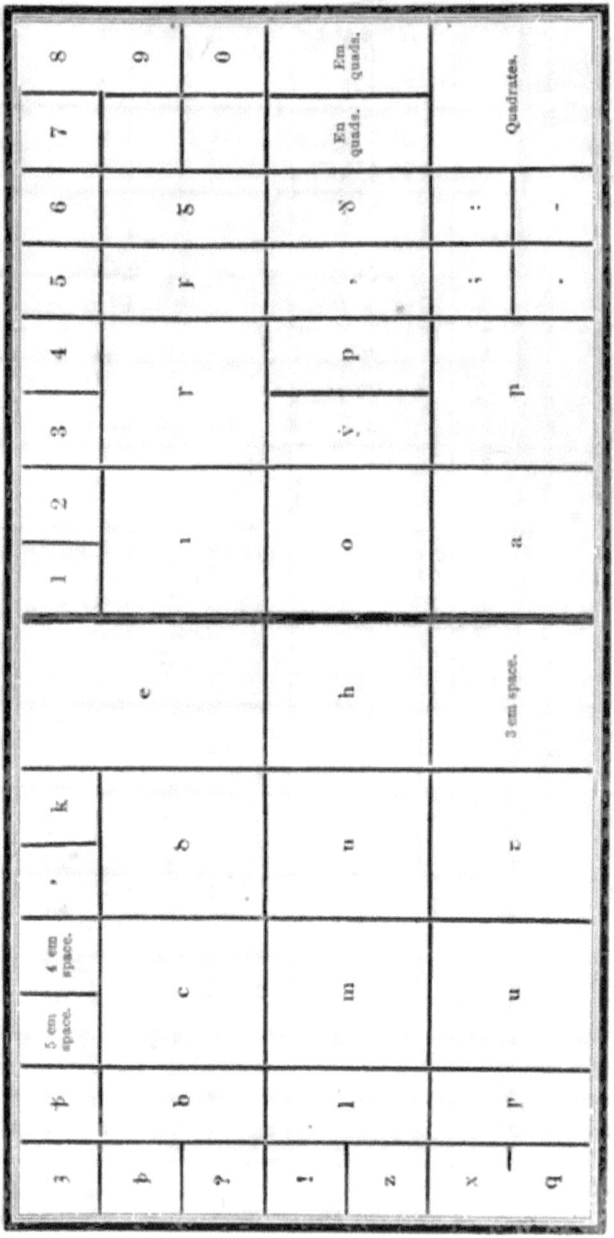

GERMAN ALPHABET.

OUTSIDE of Germany, there is perhaps no country in which German printing is so extensively executed as in the United States. We therefore present a table of German characters, with their names, and their corresponding forms in English.

German Form.	English Form.	German Name.
𝔄 ... 𝔞	A ... a	ah
𝔅 ... 𝔟	B ... b	bay
ℭ ... 𝔠	C ... c	tsay
𝔇 ... 𝔡	D ... d	day
𝔈 ... 𝔢	E ... e	a
𝔉 ... 𝔣 𝔣𝔣	F ... f ff	ef, ef-ef
𝔊 ... 𝔤	G ... g	gay
ℌ ... 𝔥 𝔠𝔥	H ... h ch	hah, tsay-hah
ℑ ... 𝔦	I ... i	e
ℑ ... 𝔧	J ... j	yot
𝔎 ... 𝔨 𝔠𝔨	K ... k ck	kah, tsay-kah
𝔏 ... 𝔩	L ... l	el
𝔐 ... 𝔪	M ... m	em
𝔑 ... 𝔫	N ... n	en
𝔒 ... 𝔬	O ... o	o
𝔓 ... 𝔭	P ... p	pay
𝔔 ... 𝔮	Q ... q	koo
ℜ ... 𝔯	R ... r	er
𝔖 ... ſ 𝔰 ſſ	S ... s s ss	es, es-es
ß ſt	sz st	es-tset, tay-tset
𝔗 ... 𝔱	T ... t	tay
𝔘 ... 𝔲	U ... u	oo
𝔙 ... 𝔳	V ... v	fōw
𝔚 ... 𝔴	W ... w	vay
𝔛 ... 𝔵	X ... x	iks
𝔜 ... 𝔶	Y ... y	ipsilon
ℨ ... 𝔷 𝔱𝔷	Z ... z tz	tset, tay-tset
𝔞̈ 𝔬̈ 𝔲̈	ae oe ue	

Several of the German letters, having a somewhat similar appearance, are liable to be mistaken one for another. To aid the learner, we give such letters together, and point out the difference.

𝔅 (B) and 𝔙 (V).

The latter is open in the middle, the former joined across.

𝕮 (C) and 𝕰 (E).

𝕰 (E) has a little stroke in the middle, projecting to the right, which 𝕮 (C) has not.

𝕲 (G) and 𝕾 (S).

These letters, being both of rather a round form, are sometimes taken one for another, particularly the 𝕲 for the 𝕾. But 𝕾 (S) has an opening above, 𝕲 (G) is closed, and has besides a perpendicular stroke within.

𝕶 (K), 𝕹 (N), 𝕽 (R).

𝕶 (K) is rounded at the top, 𝕹 (N) is open in the middle, 𝕽 (R) is united about the middle.

𝕸 (M) and 𝖂 (W).

𝕸 (M) is open at the bottom, 𝖂 (W) is closed.

𝖇 (b) and 𝖍 (h).

𝖇 (b) is perfectly closed below, 𝖍 (h) is somewhat open, and ends at the bottom, on one side, with a hair-stroke.

𝖋 (f) and ſ (s).

𝖋 (f) has a horizontal line *through* it, ſ (s) on the left side only.

𝖒 (m) and 𝖜 (w).

𝖒 (m) is entirely open at the bottom, 𝖜 (w) is partly closed.

𝖗 (r) and 𝖝 (x).

𝖝 (x) has a little hair-stroke below, on the left.

𝖛 (v) and 𝖞 (y).

𝖛 (v) is closed, 𝖞 (y) is somewhat open below, and ends with a hair-stroke.

GERMAN UPPER CASE.

*	†	‡	§	‖	¶	☞	℔	℞	@	%	%	'	°
1	2	3	4	5	6	7	$	£	2 em ⏜	3 em ⏜	⌒	⏜	⌒
8	9	0		Ä	Ö	Ü	-	—	2 em ——	3 em ———			
𝔄	𝔅	ℭ	𝔇	𝔈	𝔉	𝔊	¼	½	¾	⅛	⅜	⅝	⅞
ℌ	ℑ	𝔎	𝔏	𝔐	𝔑	𝔒	⅓	⅔					
𝔓	𝔔	ℜ	𝔖	𝔗	𝔙	𝔚							
𝔛	𝔜	ℨ	𝔘	'	]	)					ꝛ	„	

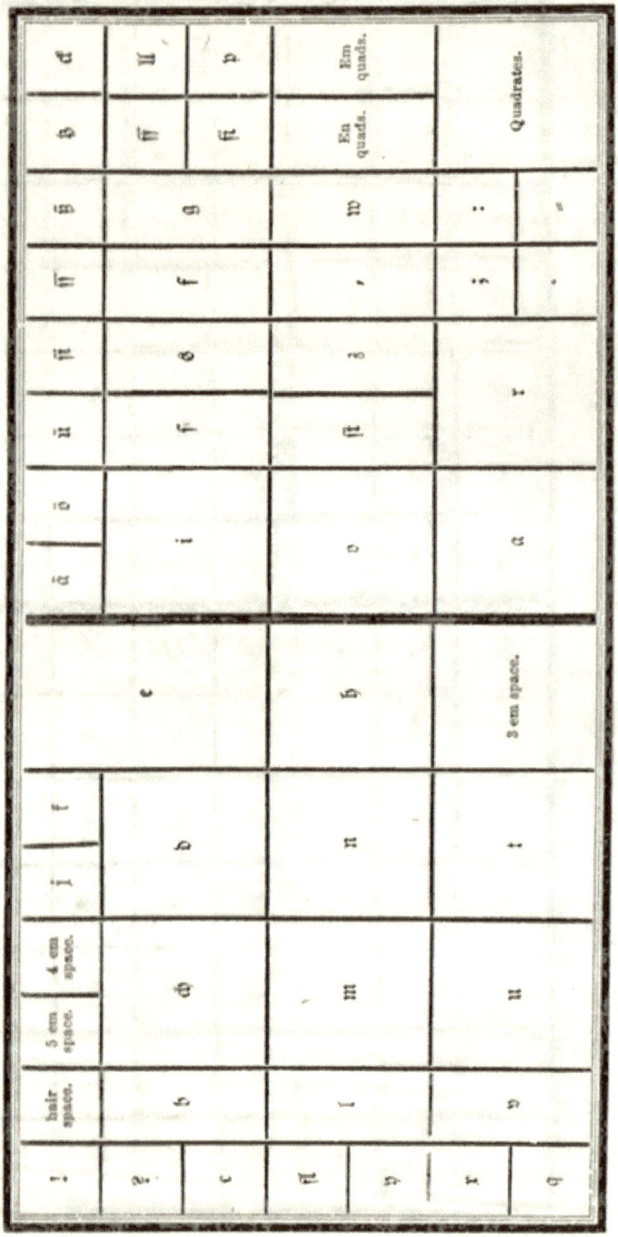

GERMAN LOWER CASE.

GREEK.

A SMALL amount of Greek types is indispensable in every considerable book printing-office. The Greek alphabet contains twenty-four letters, which we give in the following table, with the name of each character expressed in Greek and English, and its sound and numerical value.

THE GREEK ALPHABET.

Forms.	Names in Greek and English.	Sounds.	Numerical Value.
Α ... α	Ἄλφα........ Alpha	a	1
Β ... β ϐ	Βῆτα......... Beta	b	2
Γ ... γ	Γάμμα...... Gamma	g	3
Δ ... δ	Δέλτα....... Delta	d	4
Ε ... ε	Ἐψῖλόν..... Epsīlon	e short	5
Ζ ... ζ	Ζῆτα......... Zeta	z	7
Η ... η	Ἦτα.......... Eta	e long	8
Θ ... ϑ θ	Θῆτα........ Theta	th	9
Ι ... ι	Ἰῶτα......... Iōta	i	10
Κ ... κ	Κάππα...... Kappa	k c	20
Λ ... λ	Λάμβδα .... Lambda	l	30
Μ ... μ	Μῦ........... Mu	m	40
Ν ... ν	Νῦ .... Nu	n	50
Ξ ... ξ	Ξῖ............ Xi	x	60
Ο ... ο	Ὀμῖκρόν ... Omīcron	o short	70
Π ... π	Πῖ............ Pi	p	80
Ρ ... ρ	Ῥῶ.......... Rho	r	100
Σ ... σ ς	Σίγμα....... Sigma	s	200
Τ ... τ	Ταῦ.......... Tau	t	300
Υ ... υ	Ὑψῖλόν .... Upsīlon	u	400
Φ .. φ	Φῖ............ Phi	ph	500
Χ ... χ	Χῖ............ Chi	ch	600
Ψ ... ψ	Ψῖ............ Psi	ps	700
Ω ... ω	Ὠμέγα...... Omĕga	o long	800

From a desire, probably, to imitate Greek manuscript, a multitude of ligatures, abbreviations, and contractions of letters, as well as duplicates, were cast by the early type-founders. These, however, with two or three exceptions, have been quite discarded; and a fount of modern Greek is readily accommodated in a single pair of cases. The only duplicated characters in the preceding table are β and ϐ, ϑ and θ, and σ and ς. β looks best when used as an initial letter, and ϐ as a medial. ϑ and θ are used indiscriminately; but ς is employed as a final letter only.

There are twelve diphthongs or compound vowels in Greek, viz.

Six proper,—αι, αυ, ει, ευ, οι, ου; and

Six improper,—ᾳ, ῃ, ῳ, ηυ, υι, ωυ. The point under the first three letters denotes the iota, and is therefore called the *subscript iota.*

ACCENTS AND ASPIRATES.

᾿ Lenis.	ἄ Lenis acute.	῏ Circumflex lenis.
῾ Asper.	ἂ Lenis grave.	῟ Circumflex asper.
´ Acute.	ἅ Asper acute.	¨ Diæresis.
` Grave.	ἃ Asper grave.	΅ Diæresis acute.
῀ Circumflex.		῭ Diæresis grave.

Accents are nothing more than small marks which have been introduced into the language to ascertain the pronunciation of it, and facilitate it to strangers. Wherefore the ancient Greeks, to whom it was natural, never used them, as is demonstrated from Aristotle, old inscriptions, and ancient medals. It is not an easy matter to tell what time the practice of writing these accents first prevailed, though it is probable not till after the Romans began to be more curious of learning the Greek tongue and to send their children to study at Athens,—that is, about or a little before the time of Cicero.

Accents—by the Greeks called τόνοι, tones—are the rising or falling of the voice in pronouncing; which may be considered either separately in distinct syllables, or conjunctively in the same syllable.

Wherefore there are two sorts of accents: two simple, viz. the acute, ὀξὺς, figured thus [´], which denotes the elevation of the voice; and the grave, βαρὺς, shaped thus [`], to signify the falling or depression of the voice: and the circumflex, περισπώμενος, which was formed first of these two lines or points joined together thus [^], and afterward was changed into a round sort of a figure like an inverted upsilon, thus [ᴖ], but at length came to be figured like an s drawn crossway [˜]. —*Bell's Greek Grammar.*

The acute accent raises the voice, and affects one or more of the three last syllables of a word, if it has so many.

The grave depresses the voice, and affects the last syllable only.

The circumflex lengthens the sound, and affects either the last syllable of a word, or the last but one.

There are two spirits, or breathings: the asper ['], which the Greeks use instead of the letter h; and the lenis ['], which denotes the absence of the asper.

All the words that begin with a vowel have one of these breathings over them; but the vowel upsilon admits of no other than the *spiritus asper* at the beginning of a word.

In diphthongs the *spiritus* is put over the second vowel: as, *αὑτὸς*, not *ἁυτὸς*.

The letter *ρ*, at the beginning of a word, has an *asper* over it, as, *ῥέω*; and where two *ρ*s meet in a word, the first has a *lenis*, and the other an *asper*.

The apostrophe ['] is used for cutting off the vowels *α*, *ε*, *ι*, *ο*, and the diphthongs *αι* and *οι*, when they stand at the end of a word and the next word begins with a vowel: as, *παρ' αὐτῷ* for *παρὰ αὐτῷ*; *πάντ' ἔλεγον* for *πάντα ἔλεγον*.

Sometimes the apostrophe contracts two words into one: as, *κα'γὼ* for *καὶ ἐγὼ*; *ἐγῷ'μαι* for *ἐγὼ οἶμαι*; *κα'κεῖνος* for *καὶ ἐκεῖνος*.

Sometimes an apostrophe supplies the first vowel beginning a word: as, *ὦ 'γαθὲ* for *ὦ ἀγαθὲ*; *ποῦ 'ςι* for *ποῦ ἐςι*. This is chiefly used in poetry.

But the prepositions *περὶ* and *πρὸ* suffer no apostrophe though the next word begin with a vowel; for we write *περὶ υμῶν*, *πρὸ ἐμοῦ*; *περὶ αὐτον*, *πρὸ ἐτῶν*, &c.

The diæresis [¨] separates two vowels, that they may not be taken for a diphthong: thus, *ἀϋτὴ* with a diæresis makes three syllables; but without a diæresis *αυ* is a diphthong, and makes *αὐτὴ* two syllables.

Diastole [,] is put between two particles that would bear a different sense without it: thus, *ὅ,τε ὅ,τι* signify *whatever;* whereas *ὅτε* stands for *as*, and *ὅτι* for *that*. *Τό,τε* with a diastole implies *and this;* but when without, it answers to the adverb *then*.

The sign of interrogation, in Greek, is made by a semicolon [;].

The colon is made by an inverted full-point [·].

All other points are the same as in English.

The compositor will find it advantageous to bear in mind the following rules:—

1. No accent can be placed over any other than one of the last three syllables of a word.

2. The grave accent never occurs but on the last syllable; and, this being the case, the asper grave [῝] and lenis grave [῍] can be wanted only for a few monosyllables.

3. No vowel can have a spirit, or breathing, except at the beginning of a word.

4. The letter ρ is the only consonant marked by a breathing.

5. Almost every word has an accent, but very seldom has more than one; and, when this happens, it is an acute thrown back upon the last syllable from one of those words called enclitics (*leaning back*), which in that case has none, unless it be followed by another enclitic. In no other case than this can a last syllable have an acute accent, except before a full-point, colon, or note of interrogation, when the grave accent of the last syllable is changed to an acute,—a circumstance which has often led printers, who were ignorant of the reasons for accenting the same word differently in different situations, to think that there was an error in their copy, and thus to make one in their proof. Most errors, however, proceed from those who do not think at all about the matter.

PLAN OF GREEK CASES.

THE following plan of cases for Greek type is probably more convenient than any other. A Roman case may readily be altered to accommodate the lower-case sorts. Compositors who aspire to a full knowledge of their art should by all means make themselves familiar with Greek and Hebrew letters and cases.

GREEK UPPER CASE.

Α		Β		Γ		Δ		Ε		Ζ		Η															
Θ		Ι		Κ		Λ		Μ		Ν		Ξ		έ	ὲ	ε͂	ἐ	ἑ	ἐ͂	ἑ͂	ἔ	ἓ	ἒ	ἕ	ε̈	ε̈́	ε̈̀
Ο		Π		Ρ		Σ		Τ		Υ		Φ		ή	ὴ	ῆ	ἠ	ἡ	ἦ	ἧ	ἤ	ἣ	ἢ	ἥ	η̈	η̈́	η̈̀
Χ		Ψ		Ω				Kerned ᾳ		Kerned ῃ		Kerned ῳ		ί	ὶ	ῖ	ἰ	ἱ	ἶ	ἷ	ἴ	ἳ	ἲ	ἵ	ϊ	ΐ	ῒ
ᾼ		ῌ		ῼ				ᾳ		ῃ		ῳ		ό	ὸ	ο͂	ὀ	ὁ	ὀ͂	ὁ͂	ὄ	ὃ	ὂ	ὅ	ο̈	ο̈́	ο̈̀
ά	ὰ	ᾶ	ἀ	ἁ	ἆ	ἇ	ἄ	ἃ	ἂ	ἅ	α̈	α̈́	α̈̀	ύ	ὺ	ῦ	ὐ	ὑ	ὖ	ὗ	ὔ	ὓ	ὒ	ὕ	ϋ	ΰ	ῢ
´	`	῀	᾿	῾	῏	῟	῎	῝	῍	῞	¨	΅	῭	ώ	ὼ	ῶ	ὠ	ὡ	ὦ	ὧ	ὤ	ὣ	ὢ	ὥ	ω̈	ω̈́	ω̈̀

GREEK LOWER CASE.

ϰ		5 em space.	4 em space.	ϛ	ϗ	ε ʼ	σ	ς	ψ		ῥ	ῤ
β	ϐ	γ		δ			ι	η	ϑ	θ	φ	χ
κ		λ		μ		ν	ο	π	,	ρ	En quads.	Em quads.
ξ	ζ	υ		τ		3 em space.	α	ω	;	:	Quadrates.	
									.	-		

HEBREW.

THE Hebrew alphabet has twenty-two letters. Column No. 1 of the following table indicates the force of Hebrew letters when read without points. Column No. 2 gives their force when the language is printed with the Masoretic points or vowels, which are of later date than the letters. The names and numerical value of the characters are also shown.

THE HEBREW ALPHABET.

Names.	No. 1.	No. 2.	Numer. Value.
א .... Aleph	Sounded a in *war* (*vowel*)	A gentle aspirate	1
ב .... Beth	....	*Bh*	2
ג .... Gimel	g hard	*Gh*	3
ד .... Daleth	....	*Dh*	4
ה .... He	a in hate (*vow.*)	A rough aspirate	5
ו .... Vau	*u vowel*, or before a vowel, *w*	....	6
ז .... Zain	....	*Ds*	7
ח .... Cheth	....	*Hh*	8
ט .... Teth	*Th*	....	9
י .... Jod	Like *ee* in English (*vowel*)	*j* consonant, or the softer *y*	10
כ ך final.... Caph	*k* or *c* hard	....	20
ל .... Lamed	....	....	30
מ ם final... Mem	....	....	40
נ ן final..... Nun	....	....	50
ס .... Samech	....	Soft *s*	60
ע .... Ain	o long (*vowel*)	*hg*, or *hgh*, the roughest aspirate	70
פ ף final ... Phe	....	....	80
צ ץ final... Tzaddi	*j* soft	....	90
ק .... Koph	*q* or *qu*	....	100
ר .... Resch	....	....	200
ש .... Shin or Sin	....	*s* hard	300
ת .... Thau	....	....	400

LETTERS THAT HAVE A LIKENESS TO OTHERS.

Beth	Caph	
ב	כ	

Daleth	Caph	Resch
ד	ך	ר

Vau	Zain	Jod	Nun
ו	ז	י	ן

Mem	Samech
ם	ס

Gimel	Nun
ג	נ

He	Cheth	Thau
ה	ח	ת

Teth	Mem
ט	מ

Ain	Tzaddi
ע	צ

The dividing of Hebrew words not being permitted, the five following letters are cast broad to enable the compositor to justify the lines without irregular spacing:—

Aleph	He	Lamed	Mem	Thau
א	ה	ל	ם	ת

Hebrew has no capitals, and therefore letters of the same shape, but of a larger body, are used at the beginning of chapters and other parts of Hebrew works.

Hebrew reads from the right to the left, which is the case with all other Oriental languages, except Ethiopic and Armenian. In composing it, the general method is to place the nick of the letter downward, and, after putting the points to the top, to turn the line and set the points that come under the letters. If the letter has but one leg, the point is placed immediately under it; but where the letter has two legs, it is put under the centre.

The Masoretic points or vowels are subjoined under the consonant ב (beth).

1. *The Long Vowels.*				2. *The Short Vowels.*			
Kametz..........	ָ *aa*	בָ	baa	Patach.................	ַ *a*	בַ	ba
Tzeri.............	ֵ *ee*	בֵ	bee	Sœgol..................	ֶ *e*	בֶ	be
Long Chirek...	י *ii*	בִי	bii	Little Chirek........	ִ *i*	בִ	bi
Cholem..........	וֹ *oo*	בֹ בוֹ	boo	Kametz-chataph...	ָ *o*	בָ	bo
Shurek..........	וּ *uu*	בוּ	buu	Kibbutz..............	ֻ *u*	בֻ	bu

3. *Shevas, which imply a Vowel to be wanting.*

Simple Sheva..................	חְ	Chataph Patach............	חֲ	*a*
Patach furtive..................	חַ	Chataph Sœgol............	חֱ	*e*
		Chataph Kametz..........	חֳ	*o*

The last three are called compound shevas; and, in fact, they are only the short vowels, to which the simple sheva [ְ] is joined.

ACCENTS.

Hebrew accents are either mere points, or lines, or circles.

Those which are mere points or dots consist of one or two or three such points, and are always placed above the middle of the accented letter, thus,

That consisting of
- One, called *rebia*, בּ֗, i.e. *sitting over.*
- Two, called royal *zakeph katon*, ב֔, or, *the little elevator*, from its figure, which is composed of upright points.
- Three, called royal *segolta*, ב֒, an inverted [ֶ].

The lines are either upright, inclined, or transverse.

The upright is either solitary or with points or dots.

The solitary is either

- between two words, ב׀ב, termed *pesick*, or musical pause, and terminating a song.
- or under a word,
  - *Metheg*, בֽ or *bridle*, an euphonic accent at the beginning of a word.
  - Royal *silluk*, בֽ, *end*, which is placed before [:], *sophpasuk*, i.e. toward the end.

With points, namely,

- two, above the letter, royal *zakeph gadhol*, ב֕, *the great elevator*, strains the sound.
- one, below the letter, royal *tebhir*, ב֛, *broken* sound, from its figure and tone.

Inclined lines hang either above or below.

Above, toward

- the right
  - Leader *pashta*, ב֙, *extension*, extends the voice or sound, and is placed above the last letter of the word. Subservient *kadma*, ב֨, *antecedent*, to the leader *geresh;* and is placed above the penult or antepenult letter.
- the left
  - Leader *geresh*, ב֜, *expulsion*, is sung with an impelled voice.
  - *Gereshajim*, ב֞, *two expellers*, from the figure being doubled.

Below, toward

- the right —Leader *tiphcha*, ב֖, *fatigue*, from the song or note.
- the left
  - Of subservient *Merca*, ב֥, *lengthening out*, from its lengthening out the song or note.
  - *Merca kephula*, ב֦, *a double lengthening out*, from its music and figure.

The transverse line is either right or curved: thus, ־ ֮.

The right line is placed between two words, connecting them together, thus, ב־ב, and is called *maccaph*, i.e. *connection.*

The curved or waved line, ב֮, is called leader, *zarka*, or, *the disperser*, from its modulation and figure.

Circles are either entire or semi.

The entire circle is placed always above, and has a small inclined line attached to it.

Either on the left, when it is placed at the head of the word, ב֠, and is called leader *telisha* ***the greater***, or, *the great evulsion*,

Or on the right, when it is placed at the end, ב֩, and is called subservient *telisha* ***the less.***

On both together, ב֟, called leader ***karne para, the horns*** *of the heifer*, from its modulation and figure.

The semicircle is either ***solitary*** or ***pointed.***

The solitary is either *angular* or ***reflected.***

The angular is	on the right	Subservient *hillui*, ב֬, *elevated*, from the elevation of the voice.
		Munach, ב֣, *placed below*, from its position.
	on the left	Leader *jethith*, ב֚, *drawing back*, from its figure.
		Subservient *mahpach*, ב֤, *inverted*, also from its figure.
The reflected is		either single subservient *darga*, ב֧, *a degree.*
		or double, leader, *shalsheleth*, ב֓, ***a chain***, from its figure and modulation.

When joined with other points, it is either above or below the letter.

When above the letter, it has a small line attached to it on the left, ב֡, leader *paser*, *the disperser*, from the diffusion of the note.

When below the letter, it is pointed either downward, ב֑, called royal *athnach*, *respiration*, as the voice must rest upon it, and respire; or upward, ב֪, subservient, *jerah-ben-jomo*, *the* ***moon*** *of its own day*, from its figure.

PLANS OF HEBREW CASES.

The first plan shows a common case for Hebrew without points; the second exhibits a pair of cases with points.

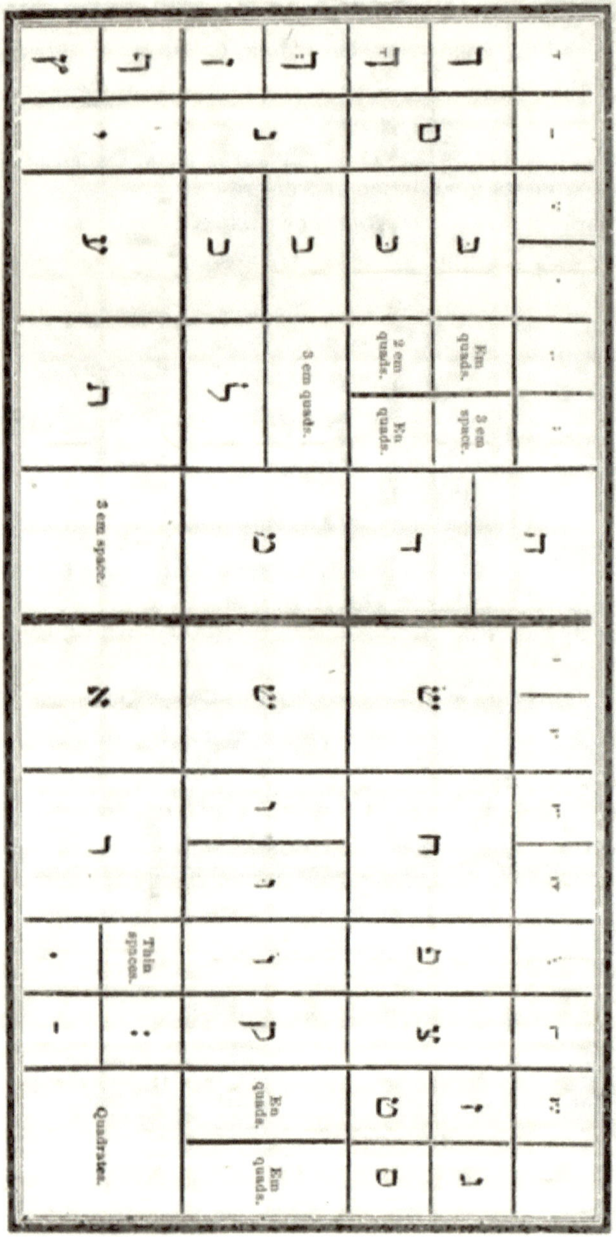

HEBREW LOWER CASE.

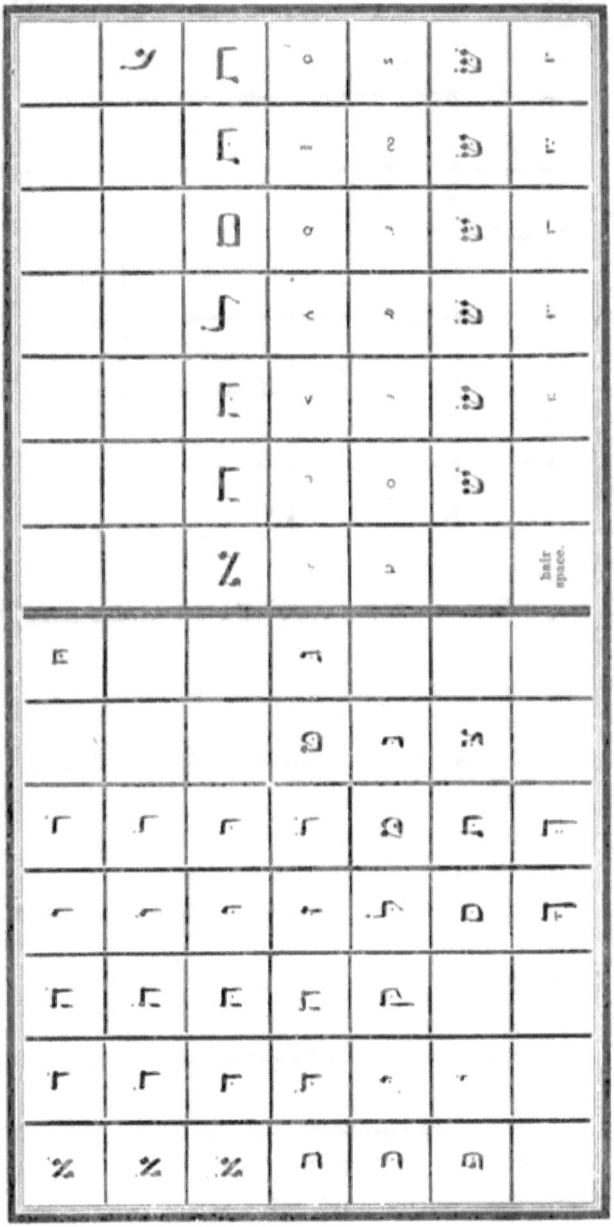

HEBREW UPPER CASE.

HEBREW LOWER CASE.

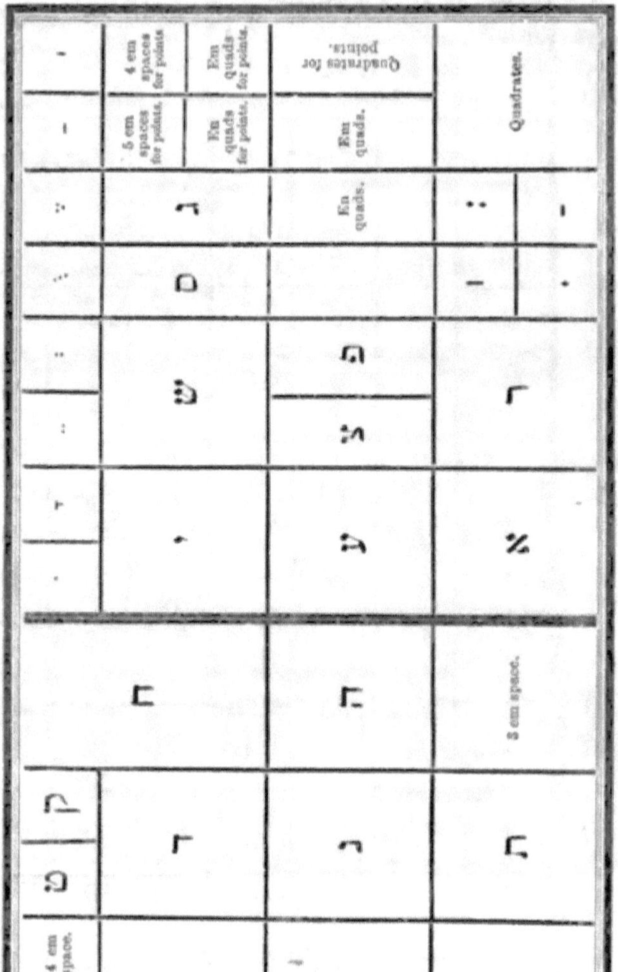

In no department of letter-founding has the progress of improvement been more decided and satisfactory than in the production of music type. The finest work of the music-stamper cannot surpass the ingenious combinations of the type-founder and printer. The music of which specimens are here given is cast on the centre of the body, and any intelligent workman may learn to compose it with facility.

EXCELSIOR.

DIAMOND No. 1.

DIAMOND No. 2.

AGATE No. 3.

MUSIC COMPOSITION.

A KNOWLEDGE of the rudiments of the art is essential to the correct and expeditious composition of music type; for, unless the compositor is acquainted with the relative time-values of the notes and rests, he cannot apportion them properly.

The manuscript copy is given to the compositor, with directions regarding the dimensions of the page required and the size of type to be employed. He counts the number of measures in the piece, and allots to each measure the amount of ems in length which the page will permit, so that there shall be a general equality of space throughout the piece.

In instrumental music, and in pieces which are not interlined with poetry, the compositor will set two or more staves simultaneously, ranging the leading notes in the under staves precisely under the corresponding ones in the upper staff; that is, a certain amount of space in each staff, in a brace, must contain the same amount of time-value. Where lines of poetry are interspersed, as in ballads and in church music, the staves are necessarily set singly; and in composing the second staff the workman must therefore constantly refer to the first, in order to make the staves correspond, proceeding in like manner with the third and fourth.

A good compositor will be careful to make the lines overlap each other, brick-wise, and not allow a joint to fall directly under another. Masters who aim to do cheap rather than good work have the music lines cast double or triple, to expedite composition. Such work has a very slovenly look, as the joints of the lines, coming under one another, are apparent in the entire depth of the staff. We have seen books set in this manner, in which all the lines seem to be composed of dotted rule, instead of a continuous stroke.

The compositor should be careful to make the stems of all the notes in a page of the same length, except those of grace-notes, which should be about half as long.

PLAN OF CASES.

THE following plan of cases is adapted to L. Johnson & Company's Diamond Music No. 1; but, modified, will serve generally for the other new styles. The figures refer to the number of the characters, as printed in the music scheme furnished with founts.

MUSIC UPPER CASE.

The figures correspond with **the number opposite the** respective characters in **the Scheme of L.** Johnson & Co.'s Diamond **Music.**

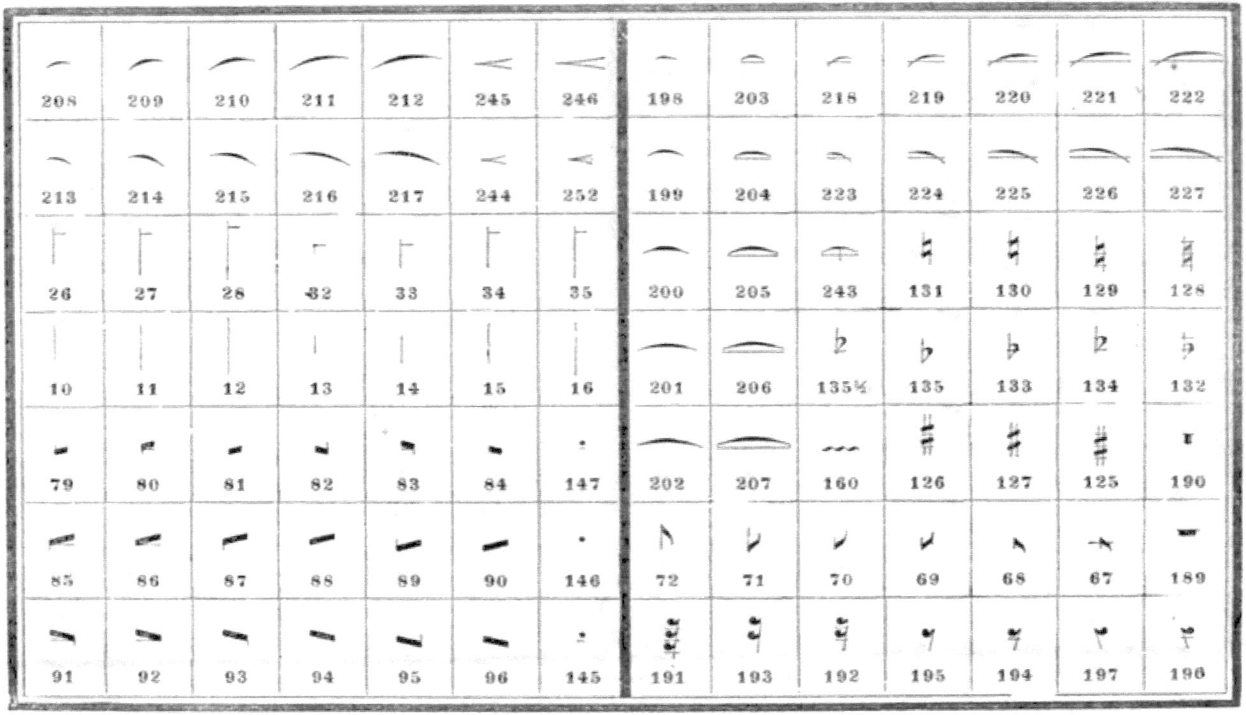

MUSIC LOWER CASE.

The figures correspond with the **number** opposite the respective characters in the Scheme of L. JOHNSON & CO.'s Diamond Music.

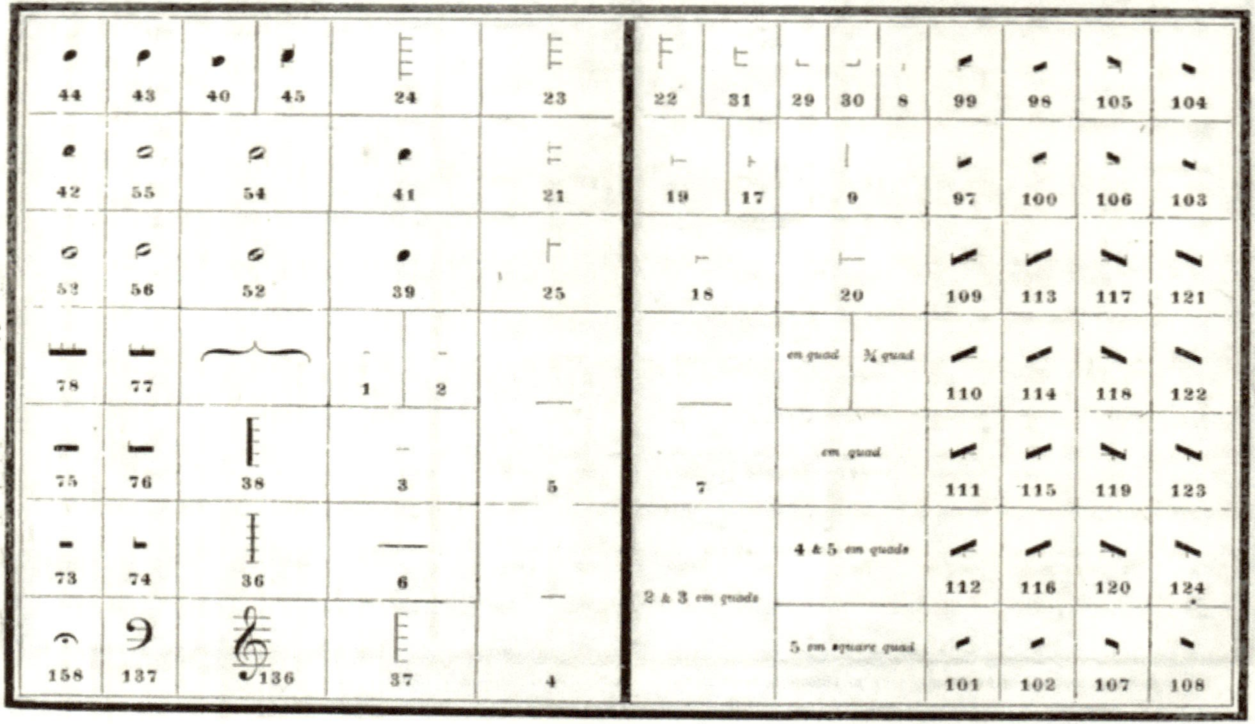

MUSIC SIDE CASE.

The figures correspond with the number opposite **the respective** characters in the Scheme **of L. Johnson & Co.**'s Diamond Music.

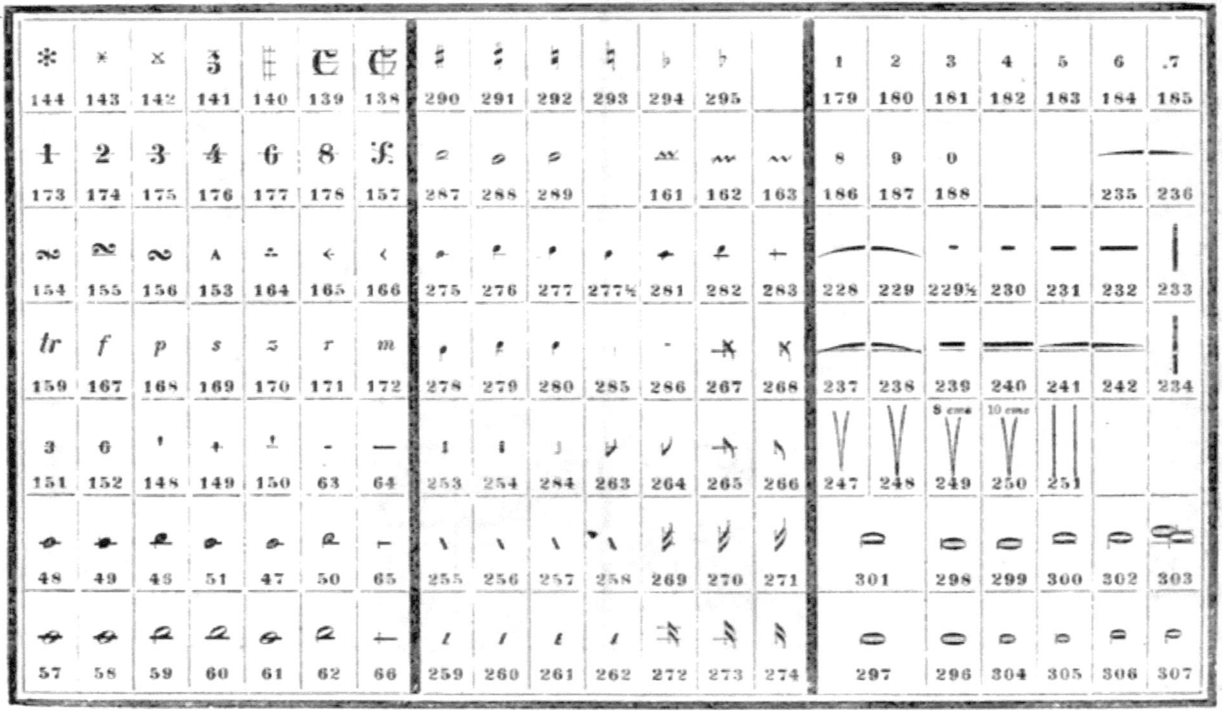

THE ART OF COMPOSITION.

GENERAL REMARKS.

EXPERIENCE proves that the apprentice foreshadows the workman, just as surely as the bend of the twig foretells the inclination of the tree. The upright, obedient, industrious lad will graduate a steady, skilful, and capable man, as unmistakably as the perverse, idling, careless boy will ripen into a lazy, dissolute, and worthless fellow. The fact is, a boy is measurably the maker of his own destiny; and if he fails to acquire a master-knowledge of the trade to which he is put, it will mainly be because he did not at his outset determine to be a master-workman. Good morals and steady industry are indispensable.

When a lad who possesses these qualities proposes to learn the art and mystery of printing, it should be inquired of him, Has he had a fair common-school education? Is he a perfect speller? Has he a turn for reading? Is his eyesight good? Is he under fifteen years of age? A true affirmative answer to all these queries will entitle him to the position of reading and errand boy. He is told the hours at which he is to come and go, and a strict punctuality is enjoined upon him. He sweeps the room,—he sorts out the pi,—he learns the position of the various letters in the case. A year spent in this way is an excellent preparative for "going to case," or learning the art of composing type.

When he is put to composition, he is told to set up one line and show it to the foreman or to the journeyman under whose care he may be placed. The errors in the line are pointed out to him, and he is required to correct them himself. When the words are perfectly correct, he justifies the line tight enough to prevent it from falling down when the composing-stick is slightly inclined, and yet sufficiently loose to enable him to lift it out with ease. In thus spacing out the line, the blanks between the words must be so graduated that, when the matter is printed, all the words will appear at equal distances apart. No matter how impatient he may be to get on, he must be drilled at this exercise till he becomes a thorough master of it. The grand doctrine to be instilled into him at first is, to do his work well and correctly; swiftness will follow as a natural consequence. He sets a second line; and after it has been made faultless he proceeds with the third, and so on till the stick is full. The utmost care must be taken to keep every letter and every line in an exact vertical position; and when he essays to empty the stick he must be taught to lift the entire mass in one square solid body, and to place it squarely and vertically on the galley. If the lines are allowed to slant either backward or sidewise, it is difficult afterward to make them stand accurately.

After the apprentice has become thoroughly conversant with the shape of every type, and can distinguish "u" from "n," "b" from "q," and "d" from "p," he is allowed to distribute type for his own use. He is taught to take up at one time no more matter than he can conveniently grasp in his left hand, which he holds so that the light falls on the face of the type, and his eye can readily read it. In distributing the various letters, he takes a word or two between the thumb and forefinger of his right hand, and the types are lightly dropped into their respective boxes.

At the outset, and as he proceeds, the novitiate must be cautioned against the acquisition of bad habits; such as swinging the body as the types are picked up, nicking the type against the stick several times before placing it in line, standing on one leg, &c.

While avoiding these ridiculous practices, a learner must acquire (if he does not possess them already) certain habitudes or principles which lie at the foundation of successful effort. The first is

Punctuality. He must conscientiously observe the time-rules of the office in coming and leaving. The early hours are the best for work; and the mind being cheered by the consciousness of doing right, the body feels the influence, and is strengthened; and when the quitting hour arrives, the amount of work accomplished will satisfy himself and his master too. The most successful masters have been distinguished for punctuality. The apprentice's time is not his own, but his master's property; and wasting it by want of punctuality, or idling during his master's absence, is simply equivalent to stealing. The second point is

Obedience. The apprentice has no right to question orders given by the master or his deputy. His duty is promptly to do as he is told, without grumbling or dissatisfaction. Let him remember that he is under orders, and that, if he ever expects to learn how to command, he must learn in his youth how to obey. He will promote his own interests by seeking to anticipate his master's wishes, and by endeavouring to make himself so useful that his services cannot well be dispensed with. Akin to this is

Courtesy. Good manners in a youth are wonderfully pleasing, and effectively aid in his advancement. Courtesy toward his master is a matter of course, and deserving of little commendation; but he must be courteous to customers when sent out on an errand, and courteous to the workmen in the office. By this means he will secure good-will, and many a friendly hint will be given to him in acquiring a knowledge of the art. The habit when fixed will bless him and others as long as he lives.

PLAN OF CASES.

THE following schemes show the order in which the letters are kept in cases in this country. In some offices, however, slight deviations will be found,—such as the transposition of the comma and w, y, p, &c.

AMERICAN UPPER CASE.

*	†	‡	§	‖	¶	☞	℔	℘	@	%	%	′	°
¼	½	¾	⅛	⅜	⅝	⅞	$	£	2 em ⏞	3 em ⏞	⏜	⏞	⏜
⅓	⅔	&	Æ	Œ	æ	œ	–	—	2 em ——	3 em ———	&	Æ	Œ
A	B	C	D	E	F	G	A	B	C	D	E	F	G
H	I	K	L	M	N	O	H	I	K	L	M	N	O
P	Q	R	S	T	V	W	P	Q	R	S	T	V	W
X	Y	Z	J	Ú	]	)	X	Y	Z	J	U	hair space.	ffl

AMERICAN LOWER CASE.

ffi	fl	5 em space.	4 em space.	'	k	e	1	2	3	4	5	6	7	8
j	b	c		d			i		s		f	g	ff	9
?													fi	0
!	l	m		n		h	o		y	p	,	w	En quads.	Em quads.
z														
x	v	u		t		3 em space.	a		r		;	:	Quadrates.	
q											.	-		

PROPOSED ***NEW*** PLAN FOR THE LOWER CASE.

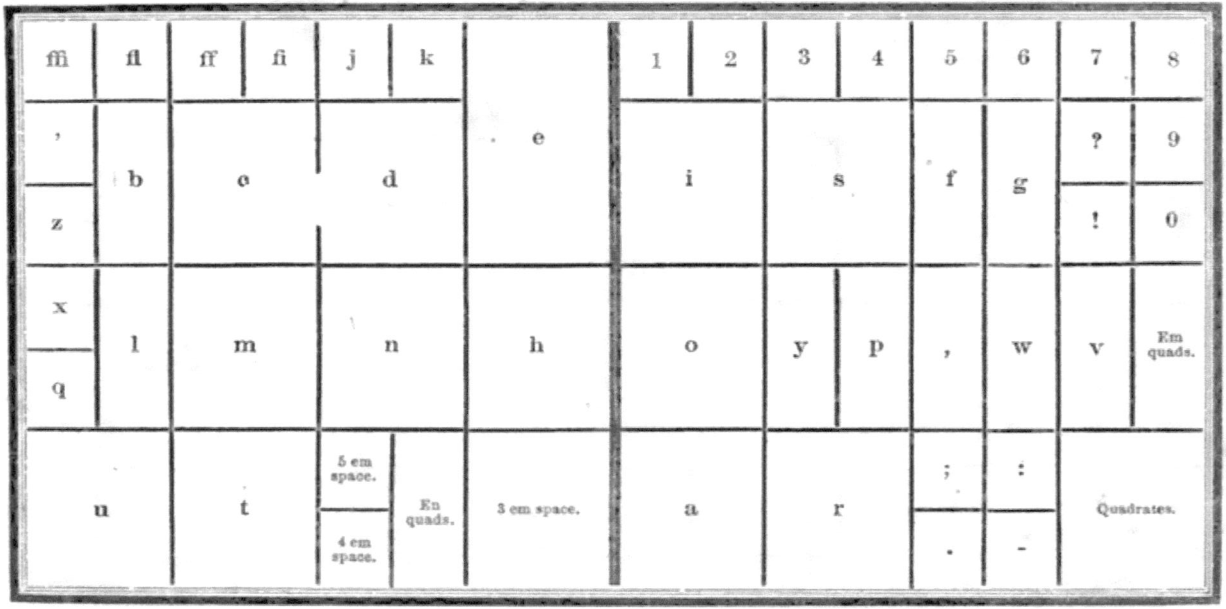

POSITION.

THE standing position of a compositor should be perfectly upright, without stiffness or restraint; the shoulders thrown back, the feet firm on the floor, heels nearly closed, and toes turned out to form an angle of about forty-five degrees. The head and body should be kept perfectly steady, except when moving from the Roman to the Italic case, the operations of distributing and composing being performed by the various motions of the arm, from the shoulder-joint alone; and if, to reach a box placed in the further part of the cases, to put in or take out a letter, he should incline the body by a slight motion, he should immediately resume his erect position. The height of a compositor and his frame should be so adjusted that his right elbow may just clear the front of the lower case by the a and r boxes, without the smallest elevation of the shoulder-joint; his breast will then be opposite the space, h, and e boxes. Sitting at work should be rarely permitted, except for lameness, weakness, old age, or other infirmity; and then the stool should be a small piece of board fastened to a single leg. Habit will render a standing position familiar and easy; perseverance in conquering a little fatigue will be amply repaid by the prevention of knock knees, round shoulders, and obstructed circulation of the blood and respiration of the lungs.

LAYING TYPE.

UNWRAP carefully the page received from the type-founder, and, laying it on a galley, soak it thoroughly with thin soap-water, to prevent the types from adhering to one another after they have been used a short time. Then, with a stout rule or reglet, lift as many lines as will make about an inch in thickness, and, placing the rule close up on one side of the bottom of the proper box, slide off the lines gently, taking care not to rub the face of the letter against the side of the box. Proceed thus with successive lines till the box is filled.

Careless compositors are prone to huddle new types together, and, grasping them up by handfuls, plunge them pell-mell into the box, rudely jostling them about to crowd more in. This is an intolerable practice.

The type left over should be kept standing on galleys, in

regular order, till the cases need replenishment. A fount of five hundred pounds of Pica may have, say, four cases allotted to it; the same amount of Nonpareil, from eight to ten cases.

DISTRIBUTING.

When a learner can infallibly distinguish from each other the letters b and q, d and p, n and u, and l and I, he may be allowed to distribute type for himself.

The head of the page being turned toward him, the learner sets a composing-rule behind the portion to be lifted, and then, placing his thumbs against the rule and his forefingers against the top line, while his remaining fingers press together both sides, he raises the matter quickly. Then, inclining sidewise his right hand, he removes the left, and allows the matter to balance momentarily in his right, while he doubles in the third finger and stretches out the thumb of the left for the reception of the matter, which he at once places in it, the rule lying as a support on the third finger, while the thumb and other fingers embrace the sides. He should take up but a few lines at a time, until he acquires facility in lifting. Large handfuls should always be avoided, as the weight is fatiguing and weakening to the wrist.

Keeping the handful in an inclined position, so that he may readily read the lines, he takes up as many letters as he can conveniently hold between his fingers,—an entire word, if practicable,—and drops the types slantingly, but with face upward, into the several boxes.

The first aim of the learner must be accuracy, even though his progress be slow. Correct distribution aids in clean composition. In time he will be able to drop his types rapidly, with hardly a glance at the boxes; and, while his fingers are flying about correctly and expeditiously, his eyes will take in the next word to be distributed; thus proceeding without cessation.

In distributing, the utmost care should be taken in placing the various spaces in their appropriate boxes. A mixing of spaces characterizes the botch.

The letter-board should always be kept clean, and the bottom as well as the face of the form well washed before it is laid on the board and unlocked; for, if any dirt remain from the ley-brush after the form is unlocked, it will sink into the matter instead of running off. This precaution taken, the pages should be well opened, and the whole form washed till the water appears to run from it in a clean state. If the form is very dirty, it is best to lock it up again and rinse the bottom of it, and proceed as before.

It is sometimes necessary to dry the letter at the fire after distributing. In this case, the type should not be used until it is perfectly cold, as very pernicious effects arise from the antimony in the composition of which the type is made. The noxious vapour which arises is sufficient warning of the effects. The compositor ought always to avoid it as a pestilence which will equally affect his respiration and his sinews, inducing lung-complaints, and causing paralysis of the hand or contraction of the fingers. Where it can be conveniently managed, it is better to distribute at night, or before meals, so that the letter may dry without artificial heat.

COMPOSING.

Composing is a term which includes several exercises, as well of the mind as the body; for, when we are said to compose, we are at the same time engaged in reading and spelling what we are composing, as well as in taking care to space and to justify our matter.

When copy is put into the hands of the compositor, he should receive directions respecting the width and length of the page; whether it is to be leaded, and with white lines between the breaks; and whether any particular method is to be followed in the punctuation and in the adoption of capitals. These instructions being given, the compositor will make his measure to the number of ems directed, which is done by laying them flatwise in the composing-stick, and then screwing it up sufficiently tight to prevent the slide from moving. He then fits a composing-rule to the measure, and, his case being supplied with letter, he commences his work.

The left hand, which contains the composing-stick, should always follow the right, which takes up the letters. If the left be kept stationary, considerable time is lost in bringing each letter to the stick, because the right hand has, consequently, to traverse a much greater space than is necessary. The eye should always precede the hand, constantly seeking for the next letter while the fingers are picking up one just selected. Each letter should be taken up by the upper end. This method will effectually prevent any false motion, and preclude the necessity of turning the letter when in the hand. If possible, a sentence of the copy should be taken at one time, and, while putting in the point and quadrate at the end of the sentence, the eye may revert to the copy for the next. It is to dexterity in these particulars that compositors are indebted for swiftness. The time thus gained is very considerable, while all appearance of bustle or fatigue is avoided. By taking a sentence into the memory at one time, the connection of the subject is preserved, and the punctuation rendered less difficult.

Those who are careful in distribution find the advantage of it in composition. Foul or slovenly workmanship is disgraceful. To avoid this, a compositor should accustom himself to glance over each line as he justifies it, and correct any error as he proceeds, which he may do with little impediment to his progress.

SPACING.

Uniformity in spacing is, unquestionably, a most important part of a compositor's occupation; this requires both care and judgment, and, therefore, cannot be too strongly impressed upon the mind of the beginner. Close spacing is as unworkmanlike as wide spacing, and neither ought to be permitted except in very narrow measures; and, frequently, even then with care it might partly be prevented. What is commonly called the thick space is the proper separator between each word; though this rule cannot always be adhered to in narrow measures when large type is used. It is not sufficient merely to have a line here and there uniformly spaced: a careful compositor will give every page that uniformity of appearance which is a chief excellency. The beginner should remember that it is better to do little, and to do that little *well*, than to put together a great number of letters without any regard to accuracy and uniformity.

Where a line is evenly spaced, and yet requires justification, the additional space should be put between those words in the line where it will be least observable: viz. a d and an h, being tall, perpendicular letters, will admit an increase of space between them, but not more than a middle and thin space to a thick-spaced line; and an additional space may be placed after a kerned letter, the beak of which may bear upon the top of an ascending letter,—as the f and the h, i, l, &c.

The same rule should be observed where it is necessary to reduce the spacing of a line, less space being required after a sloping letter than after a perpendicular one. The comma requires only a thick space, but the other points should have a hair space before and an en quadrate after them, except the full-point, which should have an em quadrate, as terminating a sentence. Should it be necessary to reduce the spacing generally, the spaces after the points must be altered in the same proportion. Spaces are now cast to such regular gradations, that the compositor can urge no reasonable excuse either for bad justification or improper spacing.

JUSTIFYING.

Accurate justification is absolutely essential, as the letters will be warped sidewise in a loose line, making it impossible to get a fair impression from the type. Besides, the letters are liable to be drawn out by the suction of the rollers, to the detriment of the form and the press. The instructor of an apprentice should occasionally pass his finger along the side of matter set by him; and if the lines should not prove evenly justified, they should be put into the composing-stick again and properly corrected.

HEAD-LINES.

Head-lines are generally set in small capitals of the same fount, or in Italic, and sometimes in capitals. Italic capitals of letter somewhat smaller than the body of the work, with folios of a proportionable size, have a neat appearance.

NOTES.

The usual rule for note-type is two sizes less than the text of the work: thus, to Pica work, Long Primer; Small Pica, Bourgeois; Long Primer, Brevier. Side-notes are usually smaller in proportion. When side-notes or references drive down below the lines of the text to which they refer, the expedient of cut-in notes must be resorted to. This is a difficult part of a compositor's business, and requires skill and patience to adjust all parts, so that every line of note and text may have proper and equal bearing. The reglet or lead between the lines of matter and the side-note must be cut with as much nicety as possible to the length of the text, as far as where the note is to run under; and, having accurately adjusted, by means of the quotations and justifiers, the situation of the first line of the note, such lead or reglet is added to the text as will make it precisely correspond in depth with the lines of note that stand on the side before turning: the remainder of the note is then set in a long measure, to correspond in width with the text, reglet, and side-note; and the page is made up with note, or the text begun again after the note is finished. In Bibles with notes and annotations, in law-books, and other works, it frequently happens that a page exhibits several of these alternate frameworks of note and text, which, if done well, display a workman's skill to great advantage.

BLANKING.

If the work is very open, consisting of heads, whites, &c., the compositor must be particularly attentive to their depth; so that though the white may be composed of different-sized quadrates, yet their ultimate depth must be equal to the regular body of the type the work is done in; otherwise the register of the work will be incomplete. The pressman cannot make the lines back if the compositor is not careful in making up his matter.

PARAGRAPHS.

The first line of a new paragraph is indented an em quadrate, of whatever type the work may be; though, when the measure is very wide, two or even three ems are preferable. By this means the paragraph is more strongly marked, the indention of an em only being scarcely perceptible in a long line. Authors vary materially in the mode of making paragraphs. Some carry the argument of a position to a great length before they relieve the attention of the reader; while others break off at almost every place that will admit only of a full-point. But the author's plan is to be followed, unless he direct otherwise. Authors should always make the beginning of a new paragraph conspicuous to the compositor, by indenting the first line of it far enough to distinguish it from the preceding line in case it should be quite full.

It is a practice too prevalent among compositors to drive out a word at the close of a paragraph, or even to divide it, in order to reap the advantage of a break-line. Part of a word, or a complete word, in a break-line, if it contain no more than three or four letters, is improper. It should be the business of the proof-reader to notice and check this irregularity.

The last line of a paragraph should not on any account begin a page, neither should the first line of a paragraph come at the bottom of a page if the work has white lines between the breaks: to prevent this, the compositor may make his page either long or short, as most convenient, always taking care that the odd and even pages back, so that the extra length or shortness of the page may escape observation.

INDEXES.

The index is generally placed at the end of the volume, and set in letter two sizes less than that of the work. It is always begun upon an uneven page. In setting an index, the subject-line should not be indented; but, if the article make more than one line, all but the first should be indented an em.

In preparing the copy of an index, care should be taken that the subject-words are ranged alphabetically, as the compositor will not transpose his matter afterward without remuneration.

Where several index-figures are used in succession, a comma is put after each folio; but, to save figures and commas, the succession of the former is noted by putting a dash between the first and last figures: thus, 4–8. Again, if an article has been collected from two pages, the folio of the second is supplied by *sq.*, or *sequente*, and by *sqq.*, or *sequientibus*, when an article is touched upon in succeeding pages. A full-point is not put after the last figures, because it is thought that their standing at the end of the line is a sufficient stop. Neither is a comma or a full-point placed to the last word of an article in a wide measure and open matter with leaders; but it is not improper to use a comma at the end of every article where the figures are put close to the matter, instead of running them to the end of the line.

TITLES.

Ornamental type may be used to good purpose in fancy jobs, and without violating any of the canons of a correct taste. The universal eye is pleased with ornament; and it is well to foster this fancy, just as we cultivate a poetical feeling, or a passion for music, or flowers, or any beautiful thing that God has made. But, as life should not be all music, or flowers, or poetry, so printing should not be all ornament. And as men whom nature puts in the fore-front of all other men are noted for an intellectual simplicity of life and style, so the title-page that heralds all the inner pages of a book should be printed in a style of elegance severe and unadorned: no fancy type, except a line of Scribe Text, or Old English,—no italics, unless perchance a single-line motto in Pearl caps,—no bold-face type, nor Antique, nor Gothic,—but plain, clear, light-faced letters that seem the embodiment of the soul of thought. All experienced printers incline to this simple style; but publishers sometimes interfere with this province of art legitimate only to a typographer, and insist on the indulgence of a taste which certainly owes no allegiance to any of the laws of beauty; and the printer or stereotyper who executes the book receives credit for a title-page which he would fain utterly repudiate.

We add a few hints which may assist the learner. 1st. Having divided the title into lines, and decided upon the size of type suitable for the principal one, begin by composing those of the second and third class, both in ascending and descending order. 2d. Avoid having two lines of equal length to follow or come in contact with each other. 3d. Catch-words should be set on a very reduced scale, and proportioned according to the strength of the preceding and succeeding lines; for bold catch-words detract from the general effect of the title. 4th. Close attention should be given to those title-pages which are acknowledged to be displayed with true taste and judgment.

Authors should endeavour to make their title-pages as concise as possible; for a crowded title can never be displayed with elegance or taste.

DEDICATIONS.

The dedication generally follows the title, and seldom exceeds one page. It should be set in capitals and small capitals, neatly displayed. The name of the person to whom the work is dedicated should always be in capitals, and the terms, Your very humble and very obedient, &c., should be in a smaller type, and the signature or name of the author in capitals of a smaller size than that in which the name of the personage is printed to whom the book is dedicated.

CONTENTS.

The contents follow the preface or introduction, and may be set either in Roman or Italic, generally two sizes smaller than the body of the work; the first line of each summary full, and the rest indented an em quadrate, with the referring figures justified at the ends of the respective lines.

PREFACES.

Formerly, the preface was uniformly set in Italic; at present Roman is used, one size larger or smaller than the body of the work. The running title to the preface is commonly set in the same manner as that of the body of the work. If the work has been printed with folios only, then the preface should be paged in like manner.

SIGNATURING.

The title, preface, &c. of a volume are commonly left till the body of the work is finished, as circumstances may arise in the course of its progress through the press which will induce the author to alter his original preface, date, &c., or the work may conclude in such a manner as to admit of their being brought in at the end, in order to fill out a sheet, and thus save both paper and press-work. For this reason, it is well to begin the first sheet of every work with signature B (or 2), leaving A (or 1) for the title-sheet.

It was formerly the custom to omit the letters J, V, and W in the list of signatures. But the greater convenience attending the use of twenty-five letters has recently induced several of our largest establishments to omit the letter J only.

ERRATA.

The errata are put immediately before the body of the work, or at the end of it. They should consist only of such corrections as are *indispensably necessary*, without noticing any defects in the punctuation, unless where the sense is perverted. It is strongly to be wished that works could be produced perfectly free from errors; but this is almost a vain hope while imperfection clings to humanity, and while every form is exposed to accident and every additional proof may be productive of fresh error.

HINTS HONOURED IN THE BREACH.

1. When you lay a fount of new type, don't open the papers carefully, and place the lines evenly with a brass rule in the cases, nick up; but show your skill by tumbling over each package rapidly, and bringing it down with a rush on the imposing-stone; then, roughly throwing the a's into a chaotic pile, grab them up by handfuls and work them well down in the appropriate box. The harder you jostle them down, the more you will get in. Proceed thus with each letter; and, if the operation has been vigorously performed, the value of the fount will have been reduced, say ten per cent.

2. While you set out one case, let your galley lie on the overheaped type of another case.

3. If a line is rather too tight to permit the last letter to get in easily, push it down hard with your rule or a quadrate. The type may be injured; but why didn't it fit in just right at first?

4. Empty your matter at a gentle inclination on the galley, and make it up at the same angle. You can bring it right afterward—perhaps—by the energetic application of mallet, shooting-stick, and planer.

5. When the case is half set out, shake up the type energetically, and do so very often. The exercise will strengthen your muscles.

6. Don't brush off the stone before you lay the matter down. If any sand happens to get under, the type will show its impression beautifully deep and clear on the face of the planer,—perhaps a whole word or two.

7. Don't plane till the form is locked up, as thus you save the trouble of the first planing. But, now that you do plane, hammer away, and show your musical ability in playing a tattoo on the form. Don't lay the planer tenderly and lovingly on the types, as if you were afraid to hurt their feelings, and gently tap it; but hold it off about a quarter or three-eighths of an inch, and then bring down the mallet with a will. Phew! how the planer will descend obedient to the stroke, and rebound again, and perhaps again. If the form is not smooth on the surface now, it is not your fault. Repeat this each time when the form is locked up, till it goes to press; and you may depend on it the impression will gain in boldness, if not in looks.

8. When correcting your numerous errors, don't trouble yourself to lift the lines carefully at the ends, but dig right into the head of the erring letter, and, resting your bodkin on the type below it, pry up the sinner: it does not matter if you demolish two or three types in the under line.

9. Wash your form energetically, and apply the ley bountifully with a good stiff wiry brush. Never mind rinsing: clean type is an old-fogy notion.

10. When the type is out of use, let it lie around promiscuously,—on a table, or board, or any place where it will be occasionally convenient to lay on it a mallet or tin basin. If one strip of matter is placed on another, room will be economized. Moreover, the under layers will be safe from dust.

11 (comprehensively). Do every thing in a loose way generally, letting matters go as they list, throwing your pi into spare boxes or secretly placing it on the letter-table or some out-of-the-way place, stealing sorts from your neighbour, overcharging time-work and extras, fishing for fat takes, &c.

12. If you observe these things faithfully and constantly, and don't lose your character and your situation, why—you do not receive your deserts.

TO THE APPRENTICE.

Aspiring apprentice, a word or two in your ear. If you desire success in any matter pertaining to this life or the coming, you must have a purpose,—a determination that, God helping you, you *will* achieve success. You may be poor, friendless, unknown,—your clothing scant, your stomach half filled,—your place may be at the foot of the ladder: no matter. Whatever your position may be, do your duty in it, stoutly and perseveringly, with your eye fixed far ahead and upward.

Keeping the purpose before you that you *will* rise, be obedient to your employer, attentive to your business, obliging to your shopmates, and courteous to strangers; and seize every opportunity to improve your heart, your mind, and your workmanship. Do every thing well,—no slighting, no hiding defects, aiming always at perfection. Watch those who are skilful, and strive to equal and excel them. Secure the friendship of all by deserving it. Allow no opportunity of rendering a service to pass without improving it, even if it cost you some labour and self-denial. Be of use to others, even if in a small way; for a time may come when they may be of service to you. A selfish man may get ahead faster than you; but selfishness is contemptible,—and you need not envy his success: when you achieve your object nobly, you will enjoy it, and be respected.

Always bear in mind that *character is capital.* To gain this, you must be so scrupulously honest that you would be as willing to put live coals in your pocket as a penny that is not yours. Never run in debt: do without what you cannot at once pay for, even though you should suffer somewhat. No matter what the amount of your earnings may be, save a portion every week, and invest it in a savings-bank of good standing: it will grow, and will stand you in good stead some day. Better temporary abstinence and constant plenty afterward, than unearned present comfort and future perpetual want. Never lie, openly or covertly, by word or action. A liar may deceive his fellows,—God and himself never. Conscious of falsity, a liar can have no self-respect; without self-respect, reputation cannot be achieved.

With a noble purpose as the end of all your actions, and with action becoming your purpose, your success is merely a question of time,—always provided you have some brain and abundant common sense.

IMPOSING, OR PREPARING FOR PRESS.

IMPOSITION

COMPREHENDS a knowledge of placing the pages so that they may regularly follow each other when printed and the sheet is folded up; and also the mode of dressing chases and the manner of making the proper margin. As many pages as are required for a whole or half sheet being made up, the compositor lays them upon the imposing-stone, placing the first page with the signature to the left hand facing him, and then proceeds according to one of the schemes on pp. 129–178. These will be found to contain every necessary imposition,—viz. folios, quartos, octavos, twelves, sixteens, eighteens, twenties, twenty-fours, thirty-twos, thirty-sixes, forties, forty-eights, sixty-fours, seventy-twos, ninety-sixes, and one hundred and twenty-eights. We also introduce schemes for imposing from the centre, by which means the blank or open pages may be thrown in the centre of the form, leaving the solid pages on the outside to act as bearers for the rollers, as well as for the better regulation of the impression.

All odd matter, for any form, should be divided into fours, eights, twelves, and sixteens, which is the groundwork of all the impositions except the eighteens, which differ from all the others; for instance, sixteens, twenty-fours, and thirty-twos are only octavos and twelves doubled, or twice doubled, and imposed in half sheets: for example, the sixteens are two octavos imposed on one side of the short cross; the twenty-fours are two twelves imposed on each side of the long cross; and a thirty-two is four octavos imposed in each quarter of the chase. Thus, a sheet may be repeatedly doubled. By this division, any form or sheet may be imposed, always bearing in mind that the first page of each class must stand to the left hand, with the foot of the page toward you. Having set down the first page, then trace the remainder according to the scheme which applies to its number; in proof of which, the standard rule for all other impositions may be adopted,—namely, *the folios of two pages, if placed properly beside each other, will make, when added together, one more than the number of pages in the sheet;* that is, in a sheet of sixteens, pages 1 and 16 coming together will add up 17, and so 9 and 8 will make 17, &c.

In half sheets, all the pages belonging to the white paper, and reiteration, are imposed in one chase. So that when a sheet of paper is printed on both sides with the same form, that sheet is cut in two in the short cross if quarto or octavo, and in the short and long cross if twelves, and folded as octavo or twelves.

TYING UP PAGES.

In tying up pages, use fine twine, winding it four or five times round it, and fastening at the right-hand corner, by thrusting a noose of it between the several turnings and the matter with the rule, and drawing it perfectly tight, taking care always to keep the end of the cord on the face of the page. While tying it, keep the forefinger of the left hand tight on the corner, to prevent the page from being drawn aside.

The twine being fastened, the compositor removes the page from the ledges of the galley, to see if the turns of cord lie about the middle of the shank of the letter; if they lie too high,—as most commonly they do,—he thrusts them lower; and if the page be not too broad, he places the fore and middle finger of his right hand on the off side of the head of the page, and his thumb on the near; then, bending his other fingers under, he presses them firmly against the head of the page; he next places the fingers of his left hand in the same position at the foot of the page, and, raising it upright, lays it on a page-paper; then, with his right hand he grasps the sides of the page and the paper, which turns up against the sides of the page, and sets it in a convenient spot under his frame, placing it on the left hand, with the foot toward him, that the other pages that are in like manner set down afterward may stand by it in an orderly succession until he comes to impose them.

If the page be a quarto, folio, or a broadside, it is, of course, too wide for his grasp; and he therefore carries the galley and page to the imposing-stone, and turns the handle of the galley toward him, and, taking hold of the handle with his right hand, he places the ball of the thumb of his left hand against the inside of the head ledge of the galley, to hold it and keep it steady, and by the handle draws the slice with the page upon it out of the galley, letting the slice rest upon the imposing-stone; he then thrusts the head end of the slice so far upon it, that the foot of the page may stand an inch or two within the outer edge of the stone, and, placing his left hand against the foot of the page, he quickly draws the slice from under the bottom of the page.

LAYING PAGES.

In taking up his pages for imposition, the compositor tightly grasps the paper on both sides of the page, in order that it may be kept firm to the bottom of the page; for if it be left slack, the letters will be liable to slip out, unless it be particularly well tied up. Having conveyed it to the stone, he next places the last two fingers of his right hand under the head of the page, but not under the page-paper at the head of it, still grasping the sides with his forefingers and thumb; he then slips his left hand so that the palm of it may turn toward the bottom, and, lifting the page upright on his right hand, with the left he removes the paper; he next grasps again the foot-end of the page with his left hand, in the same manner as the right holds the head of it, and, turning the face of the type toward him, lays it squarely and quickly down, so that the whole page may come in contact with the face of the stone at the same time.

As this method, in inexperienced or careless hands, would frequently endanger a page containing intricate matter, it will be safer to place the pages at first on good, strong, but not coarse and rough papers, and, when they are brought to the stone, instead of lifting them up as just noticed, slide them off the papers in the same manner as before directed respecting a folio page on the slice galley, being careful that no particles of dirt remain under the page.

MAKING UP FURNITURE.

HAVING ascertained that his pages are laid down right, the compositor proceeds to dress the chases, which we will suppose to be for a sheet of octavo. Accordingly he selects a good pair of chases that are fellows as well in circumference as in other respects; and, having laid them over the pages for the two different forms, he considers the largeness of the paper on which the work is to be done, and puts such gutter-sticks between page and page, and such reglets along the sides of the two crosses, as will give the book proper margins after it is bound.

To ascertain the proper distance, and to prevent wastage of furniture, he takes short pieces of furniture, or quotations, and quadrates or reglets, to fit the space between two pages; then, pushing the pages close to them, he finds the exact width of the furniture necessary, by trying the ends of various pieces, always measuring from the edge of the lines of type above the page-cords.

By observing a proper method in cutting up new furniture, the same will be serviceable for other works as well as the one for which it is intended, even though the size of the page may differ, provided it agrees with the margin of the paper. The gutters should be cut two or three lines longer than the page; the head-sticks wider; the back furniture may run nearly down to the rim of the chase, but must be level with the top of the page, which will admit of the inner head-stick running in; the difference of the outer head-stick may go over the side-stick, and the gutter will then run up between them.

The side-stick only need to be cut exact, and the furniture will completely justify.

MAKING THE MARGIN.

THE next business is to arrange the margin, so that each page may occupy one side of a leaf, and have the proper proportion of white paper left at the sides as well as at the head and foot. The page when printed should be a little higher than the middle of the leaf, and have a little more margin on the outside than in the back.

One mode of making margin is the following:—For octavos, measure and mark the width of four pages by compasses, on a sheet of paper designed for the work, beginning to measure at one extremity of the breadth of the sheet. The rest of the paper divide into four equal parts, allowing two-fourths for the width of two separate gutter-sticks; the remaining two-fourths divide again into four equal parts, and allow one-fourth for the margin along each side of the short cross, and one-fourth for the margin to each outside page. But as the thickness of the short cross adds considerably to the margin, reduce the furniture in the back accordingly, and thereby enlarge the outside margin, which requires the greatest share to allow for the unevenness of the paper itself, as well as for pressmen laying sheets uneven, when the fault is not in the paper. Having thus made the margin between the pages to the breadth of the paper, proportion the margin at the head, in the same manner, to the length, and accordingly measure and mark the length of two pages, dividing the rest into four parts, one-fourth of which is allowed for each side of the long cross, and one-fourth for the margin that runs along the foot of the two ranges of pages. The furniture on both sides of the long one must be lessened to enlarge the bottom margin, for the reason assigned for extending the side margin.

Go the same way to work in twelves, where, for the outer margin along the foot of the pages, allow the amount of two-thirds of the breadth of the head-sticks, and the same for the inner margin, that reaches from the foot of the fifth page to the centre of the groove for the points; and from the centre of that groove to the pages of the quire, or that cut off, allow half of the breadth of the head-stick. The margin along the long cross is governed by the gutter-sticks; and it is common to put as much on each side of the long cross as amounts to half the breadth of the gutter-stick, without deducting almost any thing for the long cross, since that makes allowance for the inequality of the outer margin.

Another plan, more simple, is the following:—Having laid the pages as nearly as possible in their proper places on the stone, with a suitable chase around them, fold a sheet of paper which has been wetted for the work, or one of the same size, into as many portions as there are pages in the form, and, holding the sheet thus folded on the first or left-hand page of the form, one edge even with the left-hand side of the type, place the adjoining page so that its left side may be even with the right-hand edge of the folded paper, which will leave a sufficient space between the two pages to admit the gutter-stick, which should then be selected of a proper width to suit the form in hand, as follows:—In octavos, about a Great Primer less in width than the space between the pages, as determined by the above rule; in duodecimos, about a Pica less; in sixteens, about a Long Primer; and proportionably less as the number of pages are increased. Having thus secured the proper width for the gutter-sticks, cut them somewhat longer than the page, and, holding one of them between the two pages, above the page-cord, close the pages up to it; then open the folded sheet so as to cover the two pages, and, bringing the fold in the paper exactly in the middle of the gutter-stick, secure it there with the point of a pen-knife or bodkin; the right-hand edge of the paper thus opened must be brought to the centre of the cross-bar, which determines the furniture required between it and the pages. Having thus arranged the margins for the back and fore edge of the book, proceed in like manner to regulate the head and foot margins, by bringing the near edge of the folded paper even with the bottom of the first page, and so placing the adjoining off page that its head may be barely covered by the off edge of the folded paper, which will give the required head margin. All other sections of the form must be regulated by the foregoing measurements, when the margins for the whole sheet will be found correct.

The greater the number of pages in a sheet, the smaller in proportion should the margin be: the folded paper, therefore, should lie proportionally less over the edge of the adjoining page, both for gutter and back, in a form of small pages than in one of larger dimensions. A folio may require the page to be half an inch nearer the back than the fore-edge; while a duodecimo may not require more than a Pica em.

In imposing jobs where two or more of the same size, requiring equal margins, are to be worked together, fold the paper to the size appropriate for each, and so arrange the type that the distance from the left side of one page to the left side of the adjoining one shall be exactly equal to the width of the folded paper, as before described.

Having dressed the inside of the pages, next place side and foot sticks to their outsides; being thus secured by the furniture, untie the pages, quarter after quarter, the inner page first, and then the outer, at the same time forcing the letter toward the crosses, and using every precaution to prevent the pages from hanging or leaning; and, in order to guard against accidents, when the quarter is untied, secure it with a couple of quoins.

LOCKING UP FORMS.

First, carefully examine whether the pages of each quarter are of the same length; for even the difference of a lead will cause them to hang. Test their exactness: place the ball of each thumb against the centre of the foot-stick, raising it a little with the pressure, and, if the ends of both pages rise equally with the stick, it is a proof they will not bind; then fit quoins between the side and foot-stick of each quarter and the chase. After pushing the quoins as far as possible with the fingers, make use of the mallet and shooting-stick, and gently drive the quoins along the foot-sticks first, and then those along the side-sticks, taking care to use an equal force in the strokes, and to drive the quoins far enough up the shoulders of the side and foot sticks, that the letter may neither belly out nor hang, and the lines be kept straight and even. Quoins should be slanted on one side only, but the edges should not be bevelled. The several quarters of the form should be partially tightened before either quarter is finally locked up; otherwise the cross-bar may be sprung.

Before locking up the form, plane the pages gently over all the face. If this be properly done, a second planing is hardly necessary, provided the justification is perfect and the pages are all of the same length. But, as this is seldom the case, the second planing can hardly be dispensed with.

It often occurs that the quoins, when locked up wet, stick so tight to the furniture as to render it troublesome to unlock them: in such cases, drive the quoin up a little, and it afterward unlocks with ease.

Before lifting a form after it is locked up, raise it gently a short distance, and look under it, to ascertain whether any type are disposed to drop out. If all is right, carry it to the proof-press, and pull a good proof. Then rub it over gently with a ley-brush, rinse it well, and place it in a rack, and deliver the proof, with the copy, to the proof-reader.

MEMORANDA.

EACH part of the furniture should be in one piece, where it is practicable,—as, for instance, the gutters, the backs, and the heads; but sometimes pieces will be wanted of a width that is not equal to any regular size, and then two must be used.

All the gutters of one sheet should be cut of a precise length; so also with the backs and the heads; but each sort should be of a different length from that of the others: thus they would be easily distinguished from each other, and mistakes would be prevented.

The sheet being imposed, the stone should be cleared; the saw and saw-block put in their places, the shears, the mallet, planer, and shooting-stick, the surplus furniture, the leads, the quoins, and every other article. The compositor will tie up his page-cords, and, if he has any companions, will return to them their proportion.

The chase and furniture of one form should always be used for a similar form; that is, the chase and furniture of the outer form should be again used for an outer form, and the chase and furniture of the inner form should be again used for an inner form; they should also be put round the pages in the same order in which they were put about those of the preceding forms. For want of care or thought in these apparently trifling circumstances, trouble, inconvenience, and loss of time frequently occur; for the register will be almost sure to be wrong when this is neglected, and then the forms must be unlocked and the leads changed, to correct the fault.

Abstract Title-Deeds of Estates.

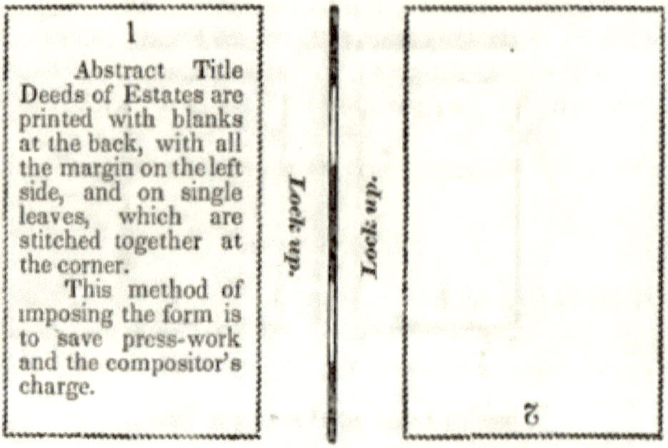

A Single Sheet of Folio.

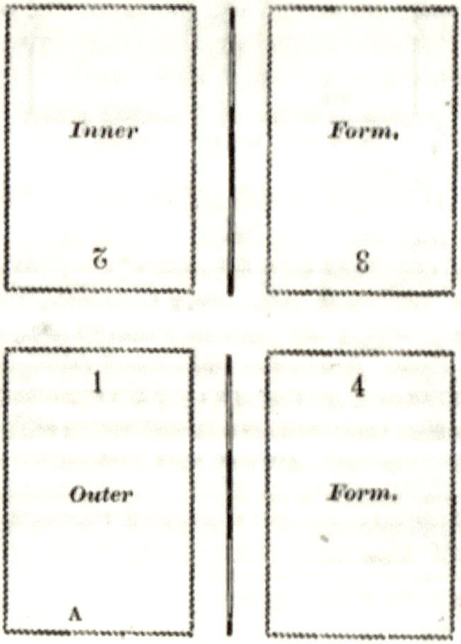

Two Sheets of Folio, Quired, or lying one in another.

Outer Form of the Outer Sheet.

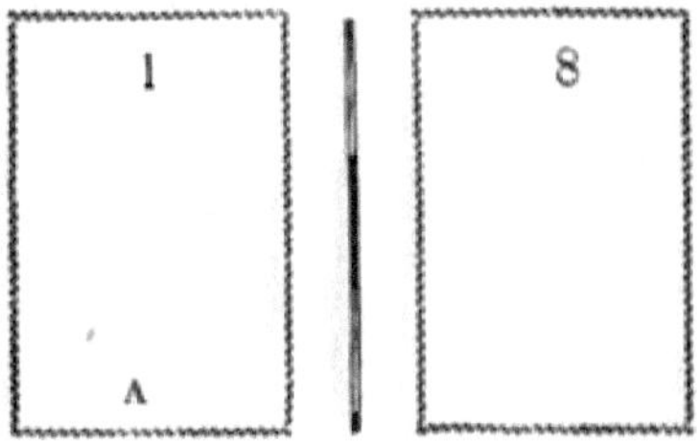

Outer Form of the Inner Sheet.

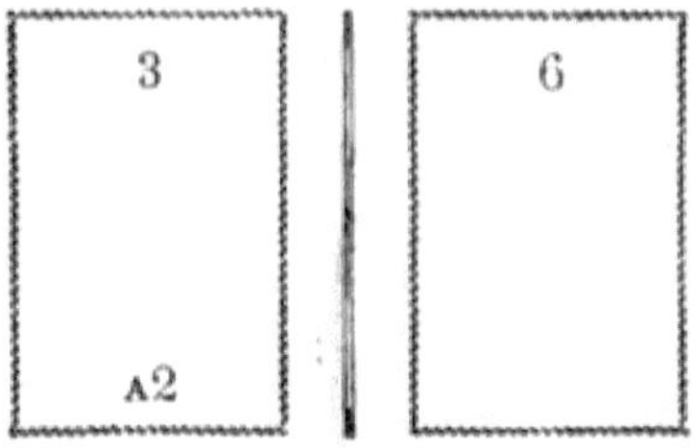

Imposing in quires may be carried to any extent, by observing the following rule:—first ascertain the number of pages, then divide them into so many sheets of folio, and commence laying down the first two and last two, which form the first sheet, and so on to the centre one, always remembering that the odd pages stand on the left and the even on the right; the folios of each two forming one more than the number of pages in the work: for example, let us suppose the work to consist of thirty-six pages, which is nine sheets of folio, then they should be laid down according to the scheme at the foot of the opposite page.

Two Sheets of Folio, Quired, or lying one in another.

Inner Form of the Outer Sheet.

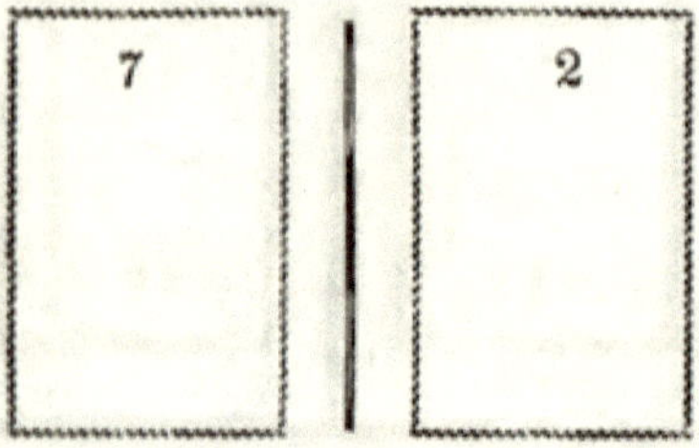

Inner Form of the Inner Sheet.

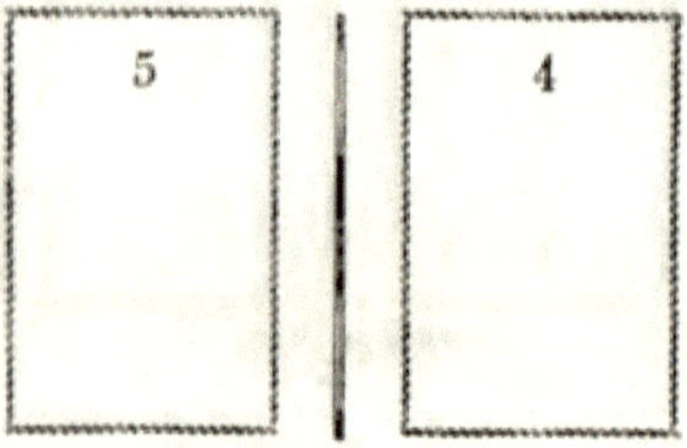

Outer.	Inner.	Sheet.	Outer.	Inner.	Sheet.	Outer.	Inner.	Sheet.
1 36	35 2,	*1st*	3 34	33 4,	*2d*	5 32	31 6,	*3d*
7 30	29 8,	*4th*	9 28	27 10,	*5th*	11 26	25 12,	*6th*
13 24	23 14,	*7th*	15 22	21 16,	*8th*	17 20	19 18,	*9th*

The furniture must be reduced in the backs of the inner sheets, to allow for stitching.

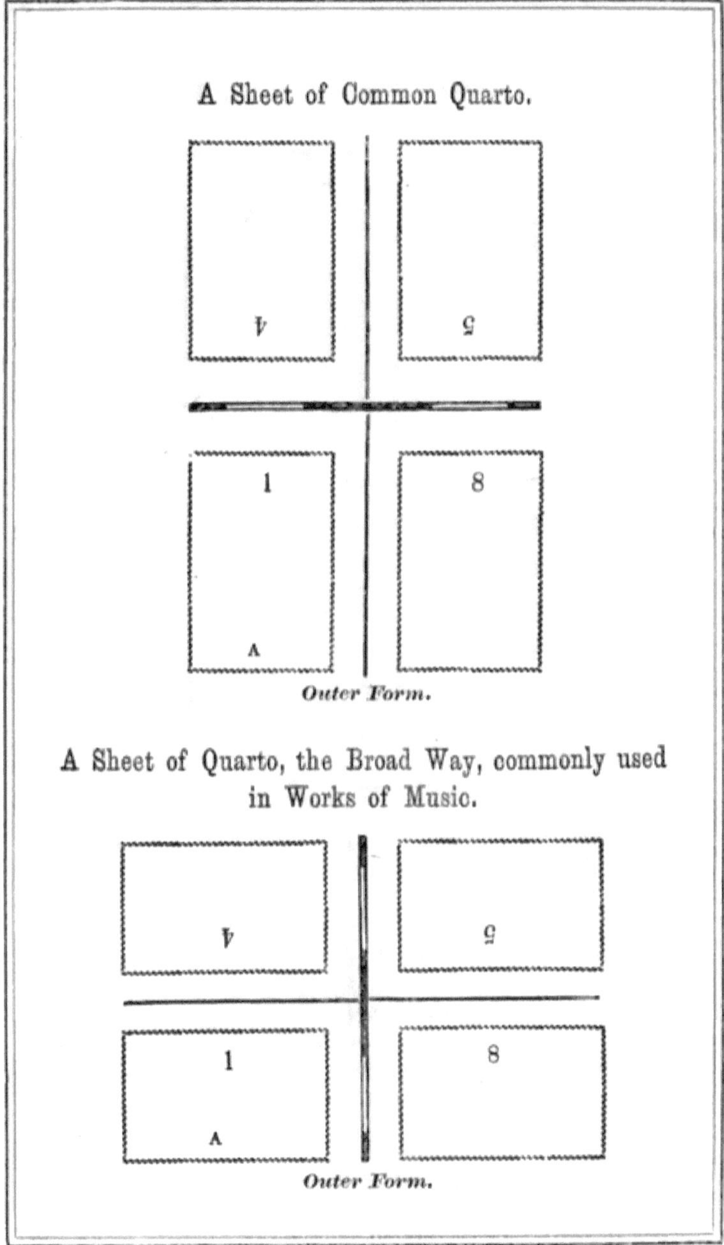

A Sheet of Common Quarto.

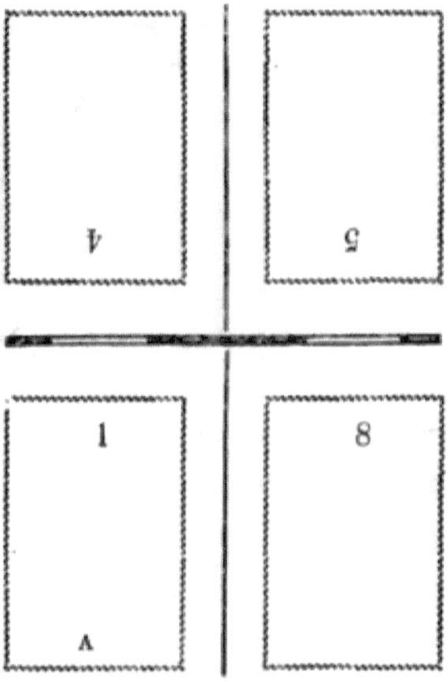

Outer Form.

A Sheet of Quarto, the Broad Way, commonly used in Works of Music.

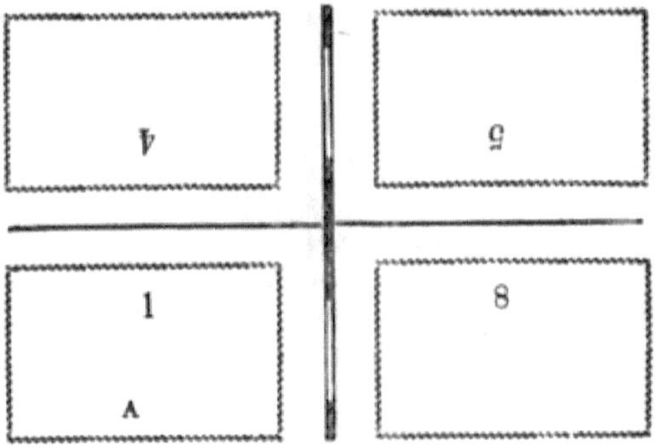

Outer Form.

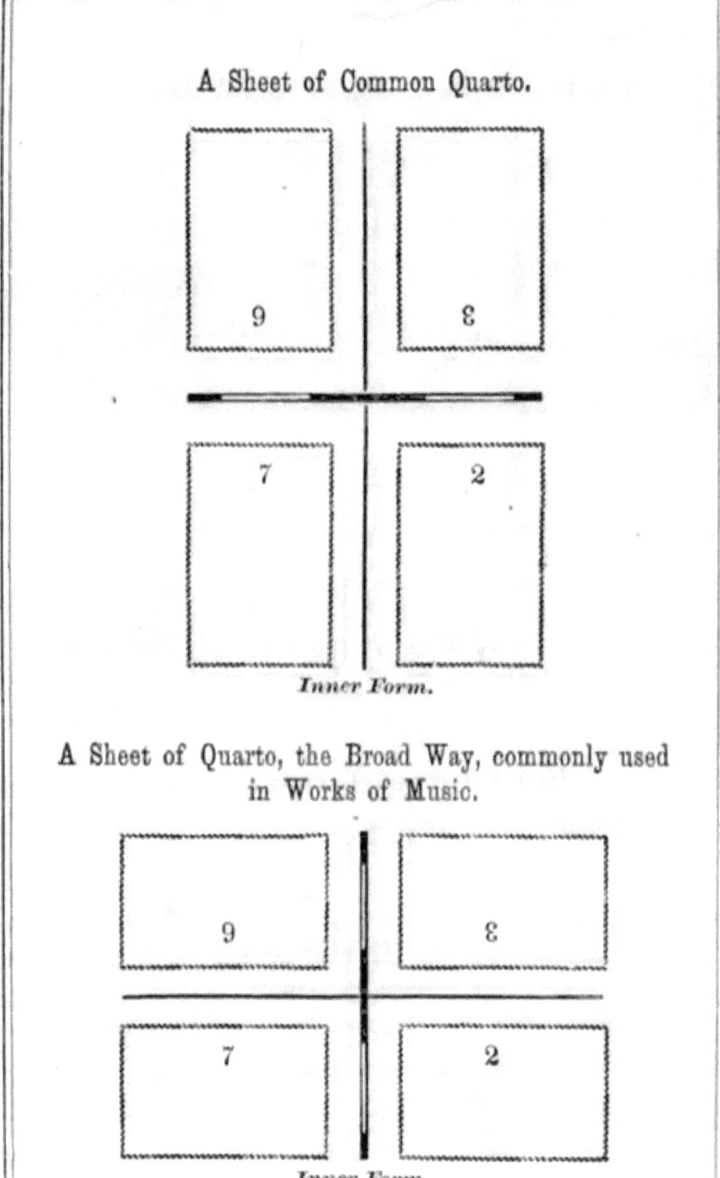

A Sheet of Common Quarto.

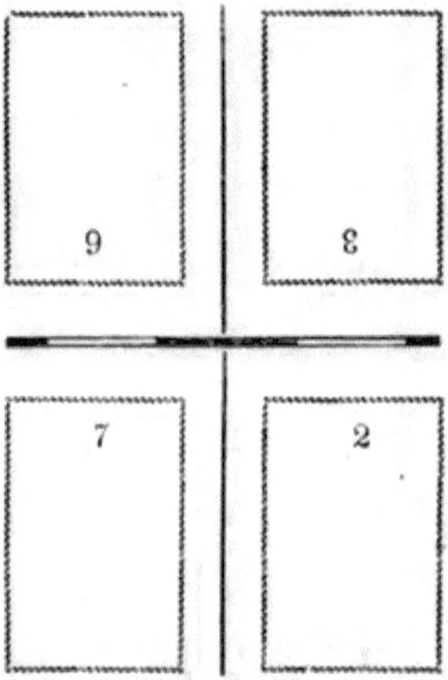

Inner Form.

A Sheet of Quarto, the Broad Way, commonly used in Works of Music.

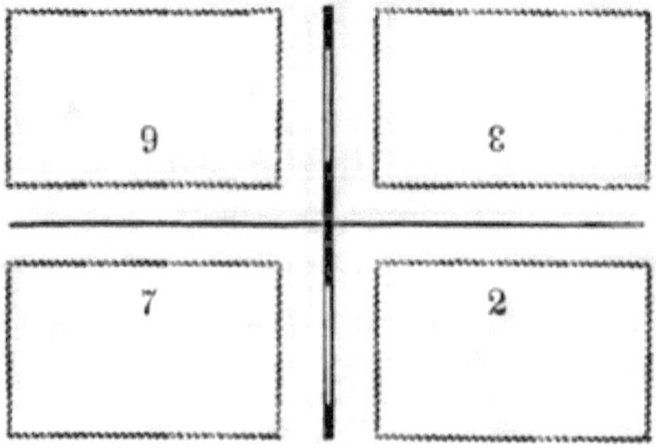

Inner Form.

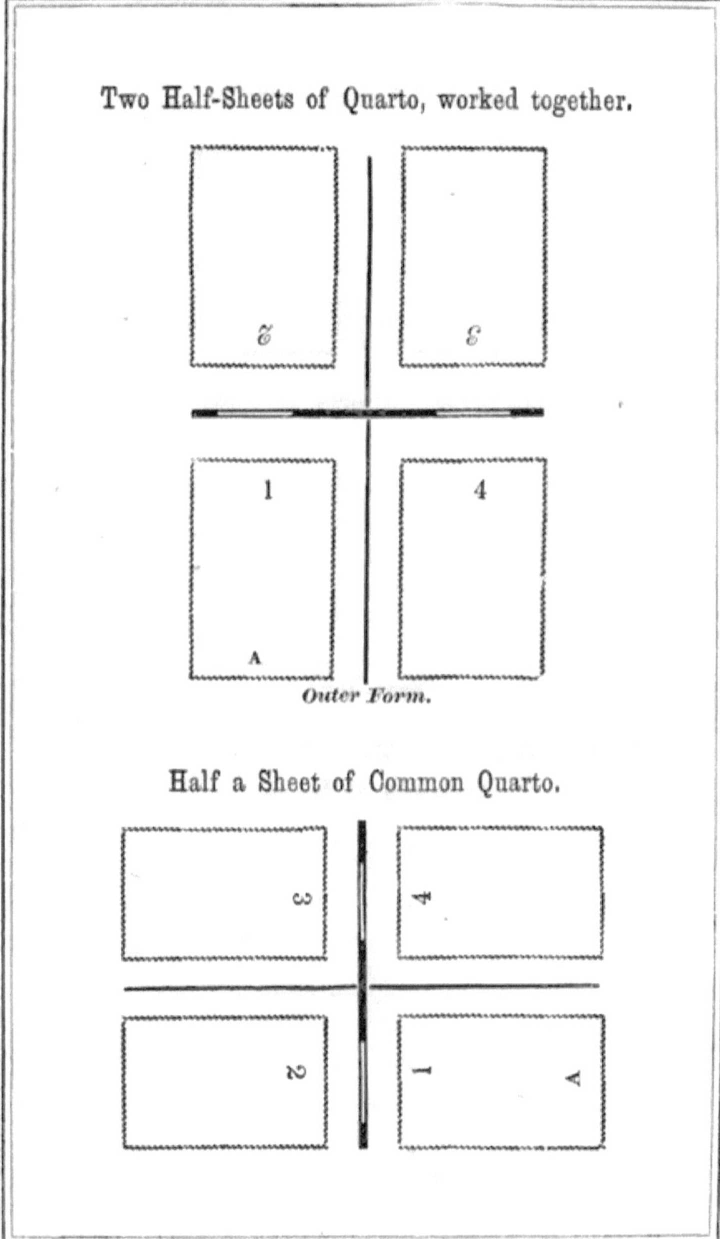

Two Half-Sheets of Quarto, worked together.

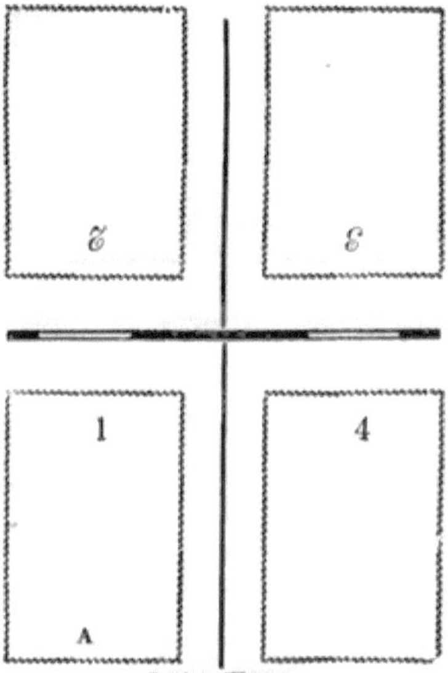

Outer Form.

Half a Sheet of Common Quarto.

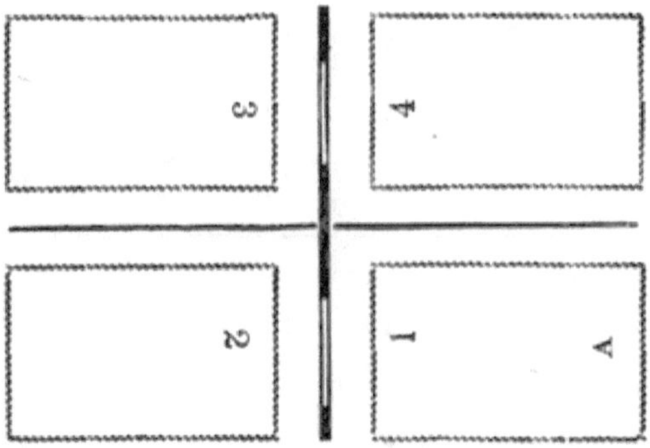

Two Half Sheets of Quarto, worked together.

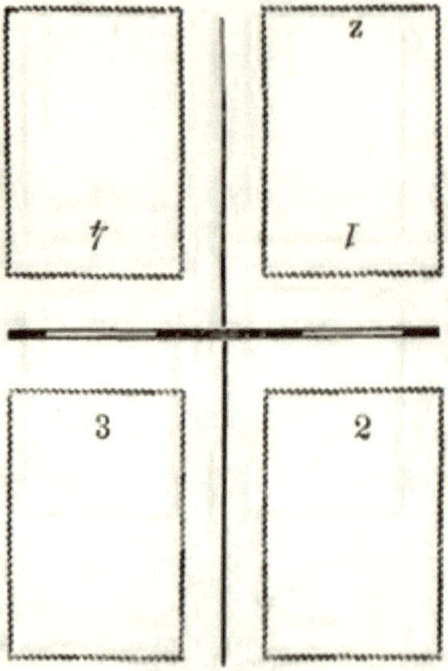

Inner Form.

Half a Sheet of Quarto, the Broad Way.

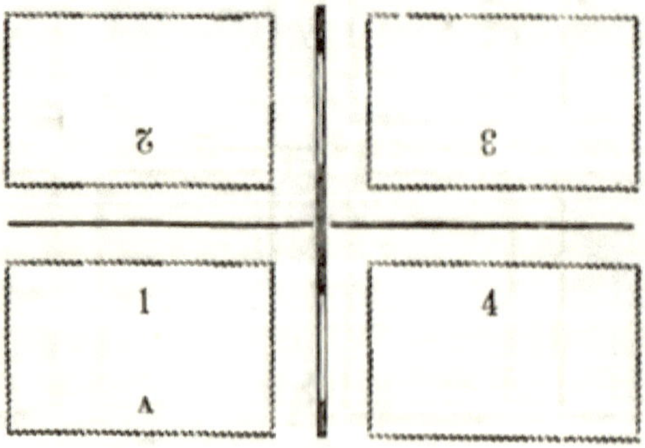

Outer Form of a Sheet of Common Octavo.

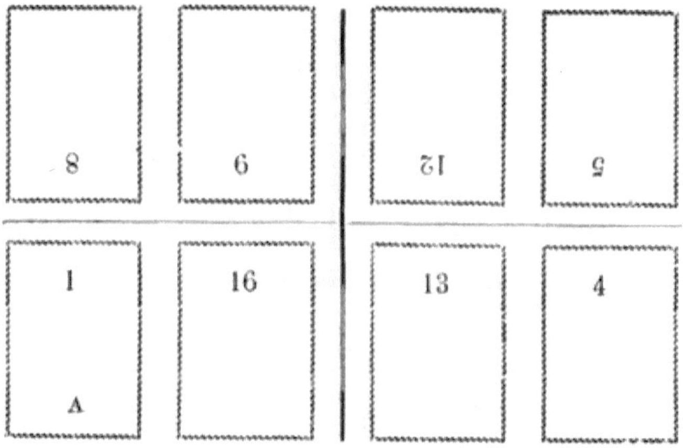

Outer Form of a Sheet of Octavo, the Broad Way.

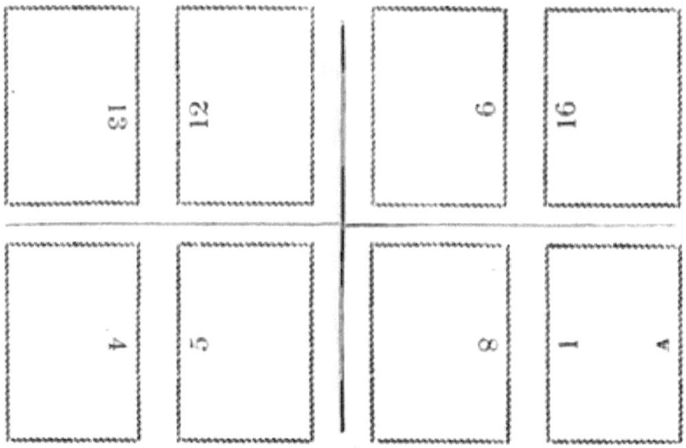

Inner Form of a Sheet of Common Octavo.

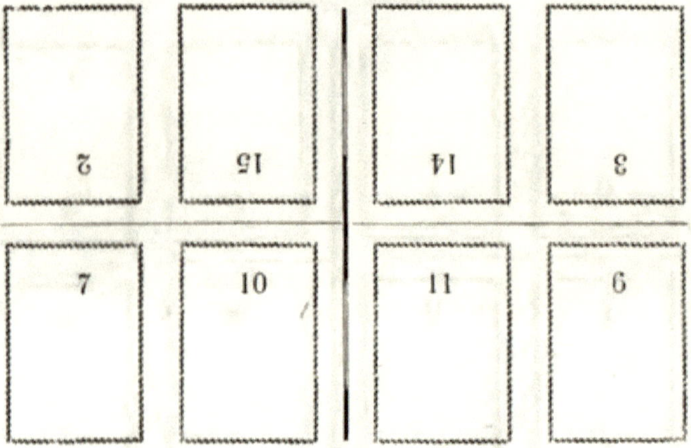

Inner Form of a Sheet of Octavo, the Broad Way.

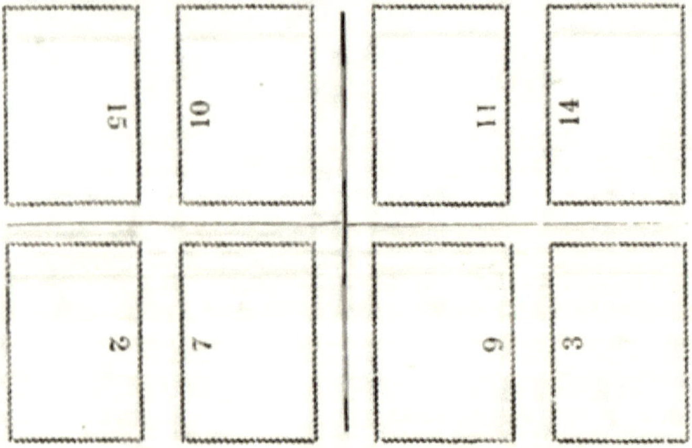

Outer Form of Two Half-Sheets of Common Octavo, worked together.

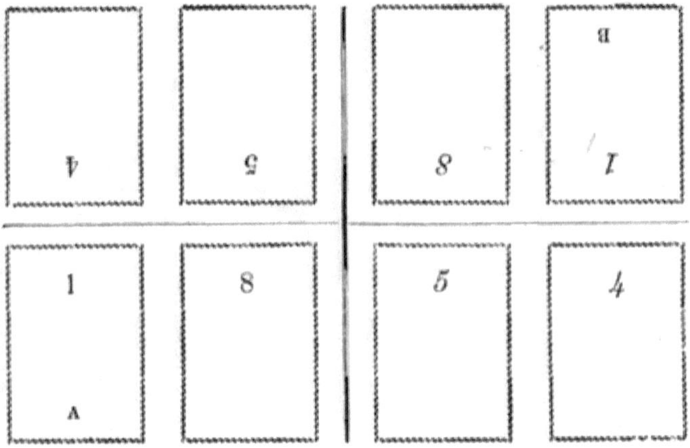

Half a Sheet of Common Octavo.

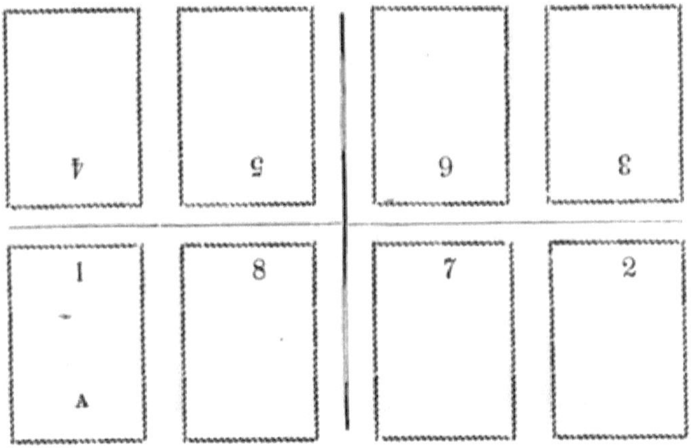

Inner Form of Two Half-Sheets of Common Octavo, worked together.

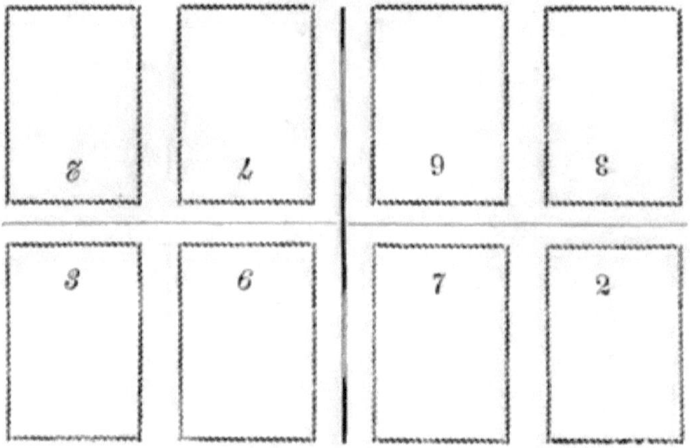

Two Quarters of a Sheet of Octavo, worked together.

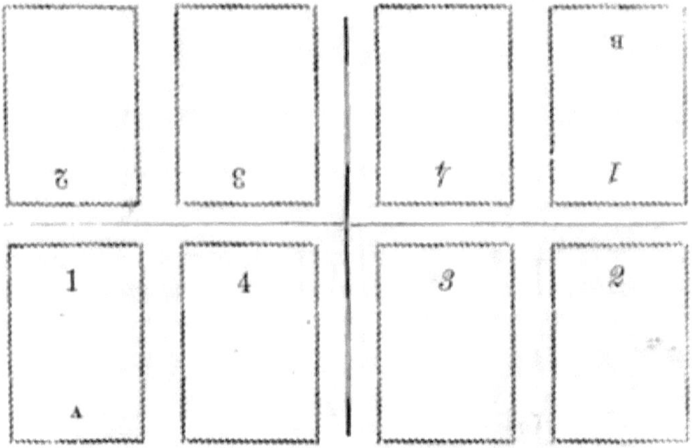

Outer Form of a Sheet of Octavo, 12 of the Work, and 4 of other Matter.

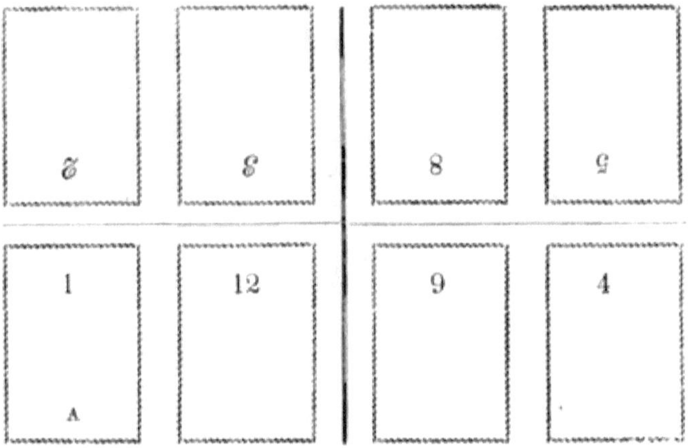

Outer Form of a Sheet of Octavo, of Hebrew Work.

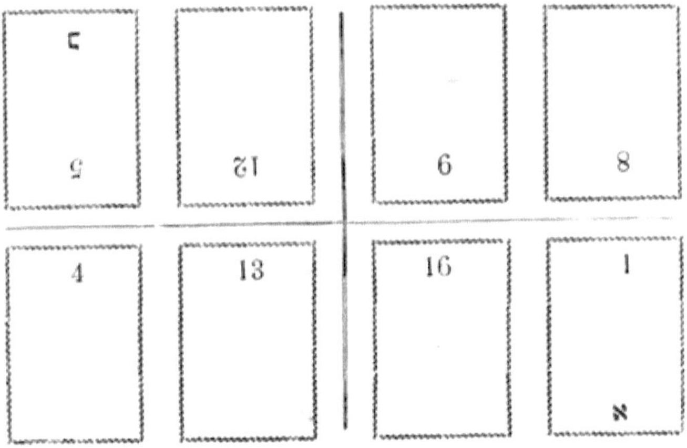

Inner Form of a Sheet of Octavo, 12 of the Work, and 4 of other Matter.

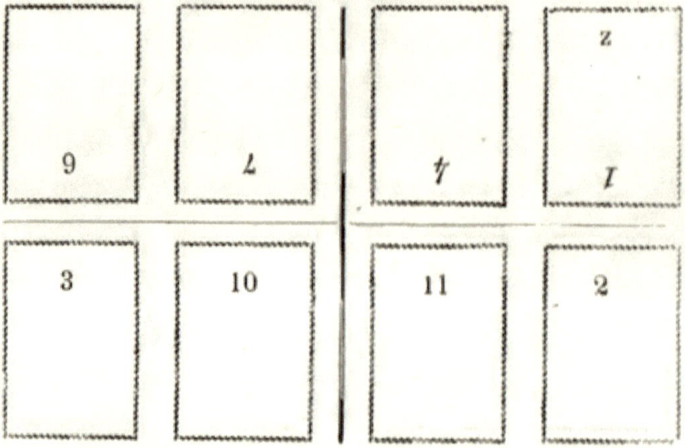

Inner Form of a Sheet of Octavo, of Hebrew Work.

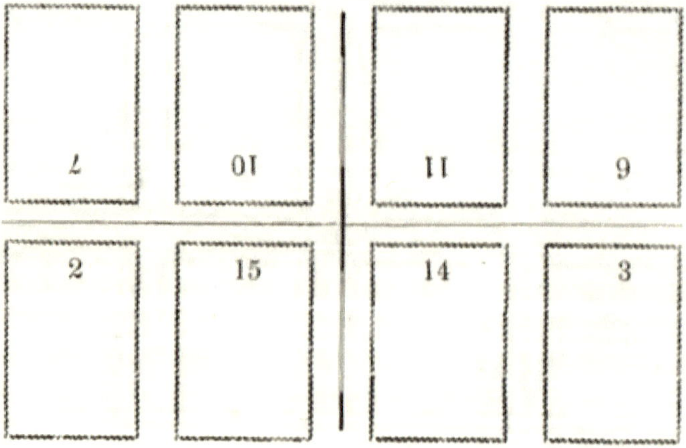

Outer Form of a Sheet of Octavo, Imposed from the Centre.

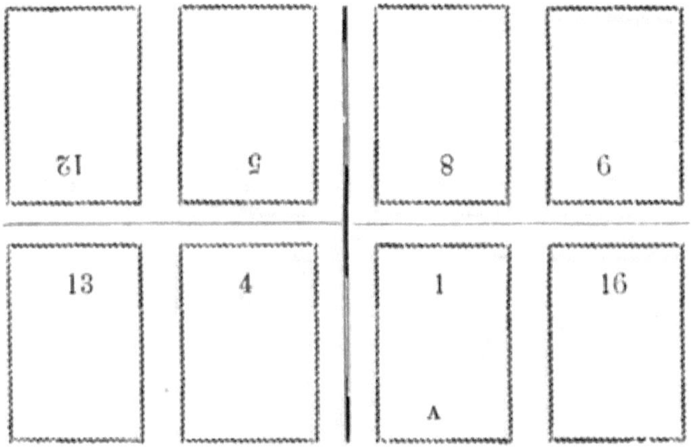

A Half-Sheet of Octavo, Imposed from the Centre.

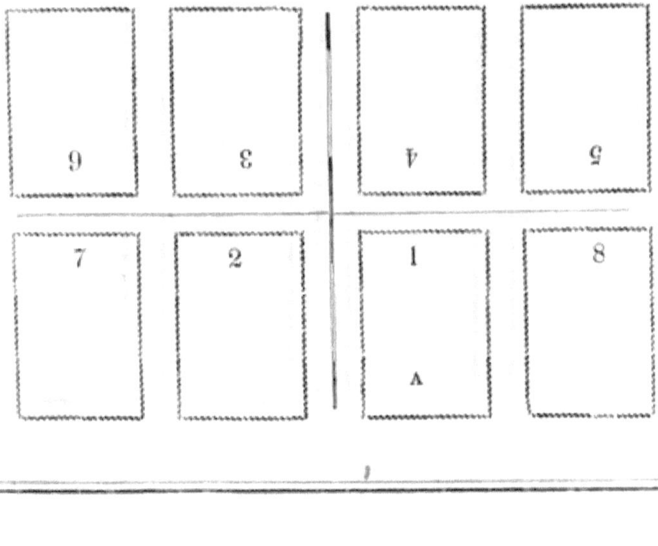

Inner Form of a Sheet of Octavo, Imposed from the Centre.

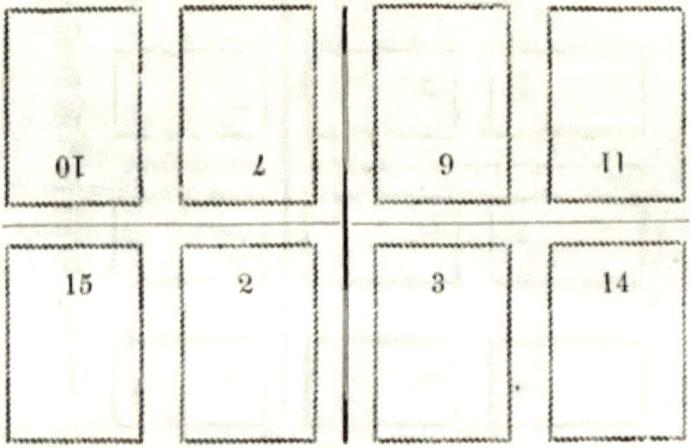

Two Quarters of a Sheet of Octavo, Imposed from the Centre.

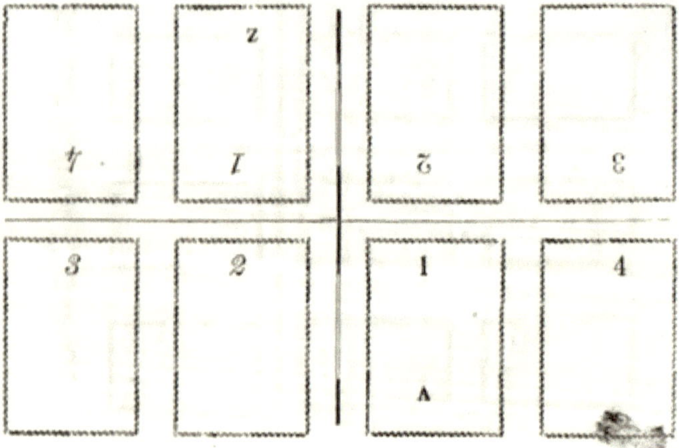

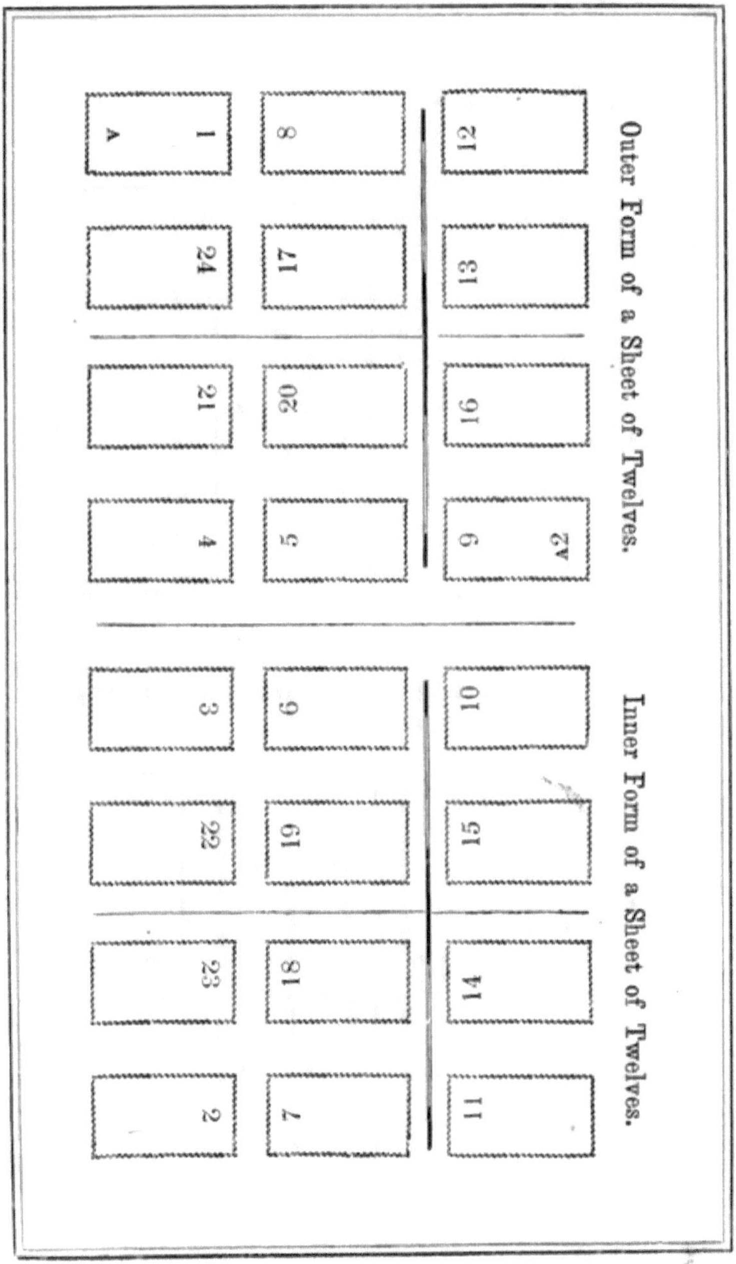

Outer Form of a Sheet of Twelves.
Inner Form of a Sheet of Twelves.

A Sheet of Twelves, without cutting.

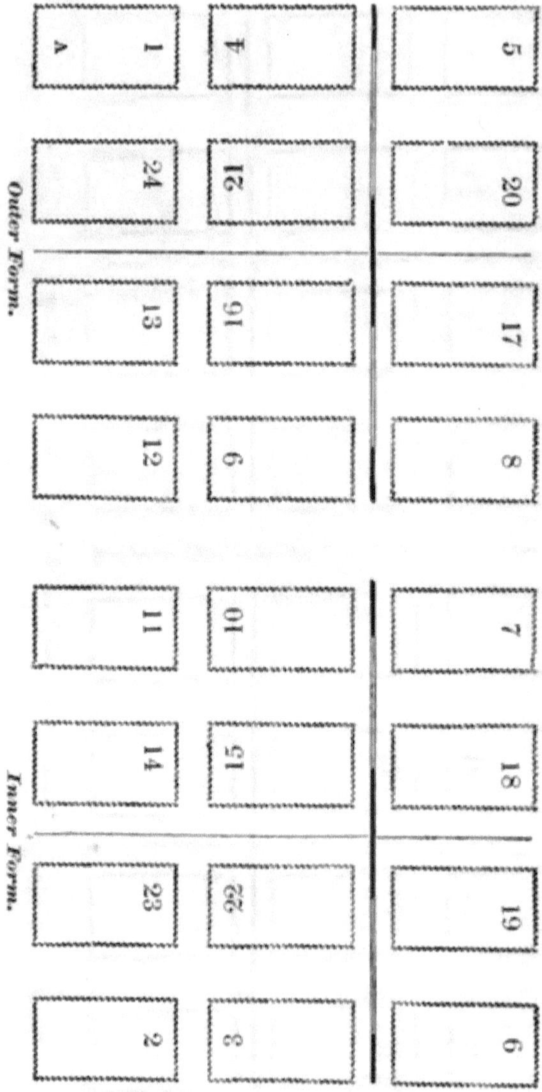

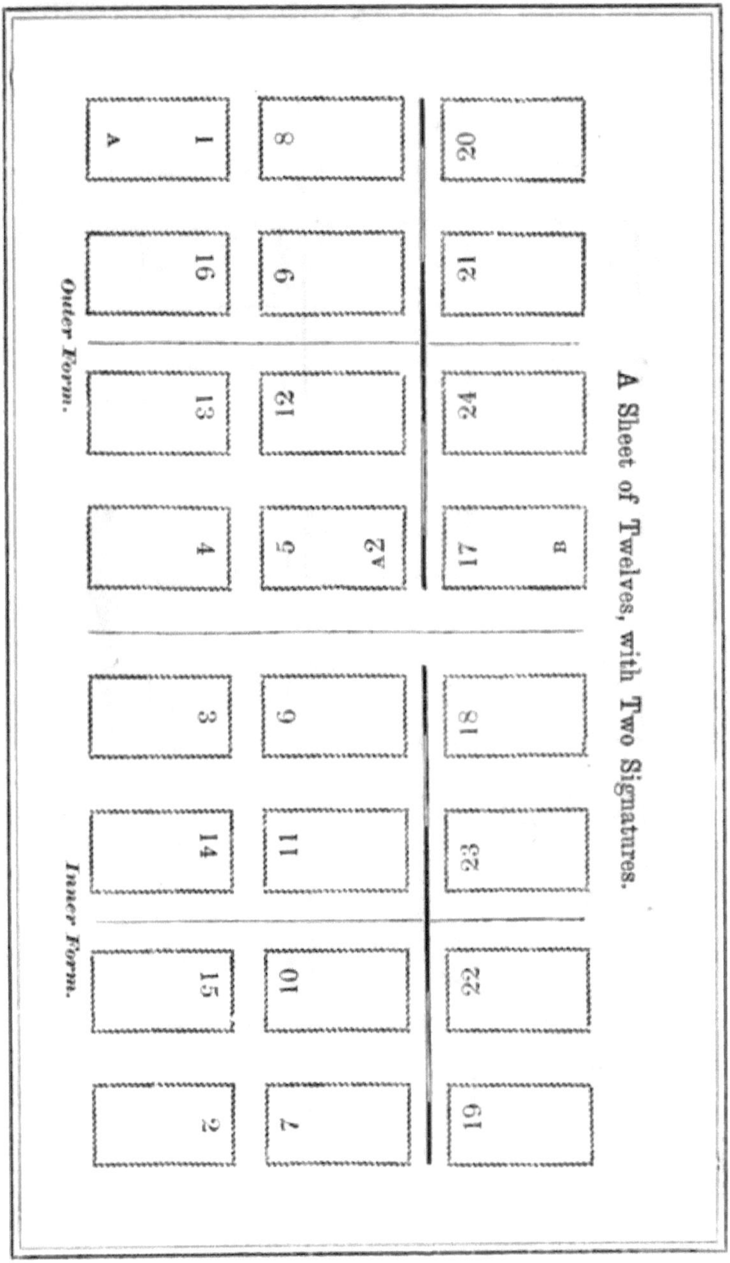

A Sheet of Twelves, with Two Signatures.

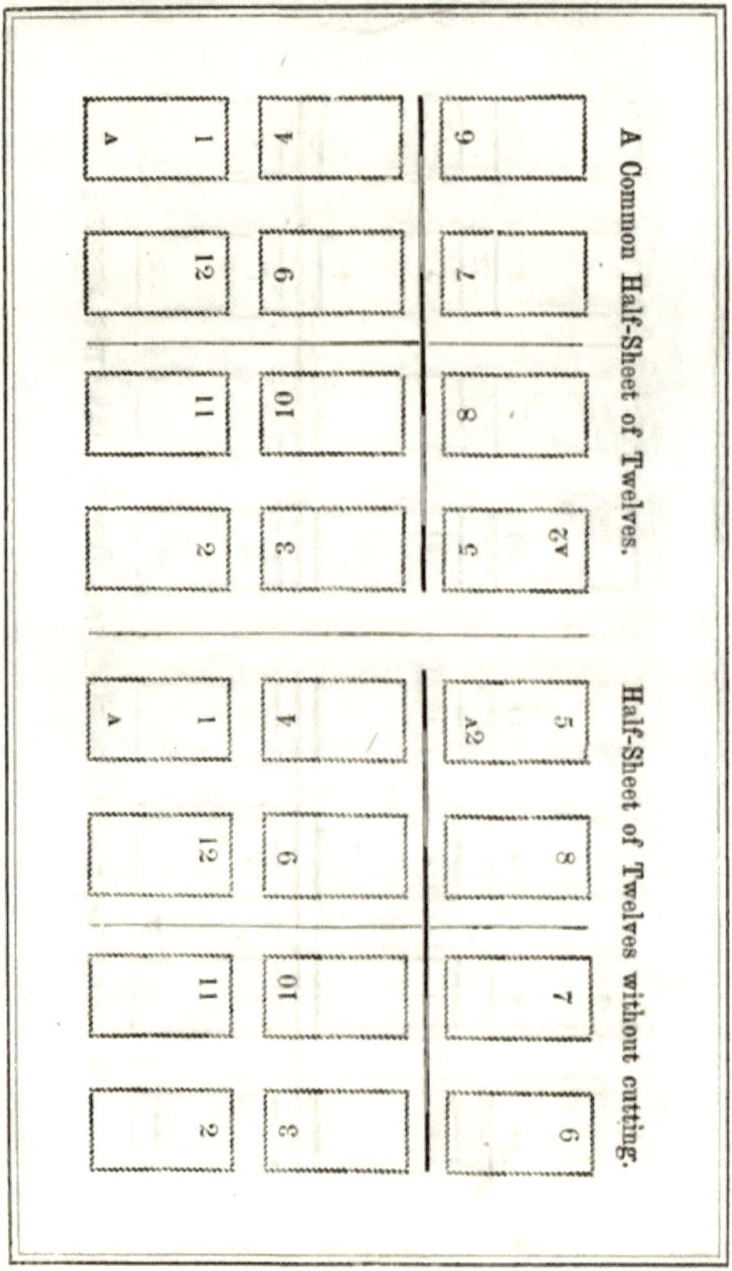

A Common Half-Sheet of Twelves.

Half-Sheet of Twelves without cutting.

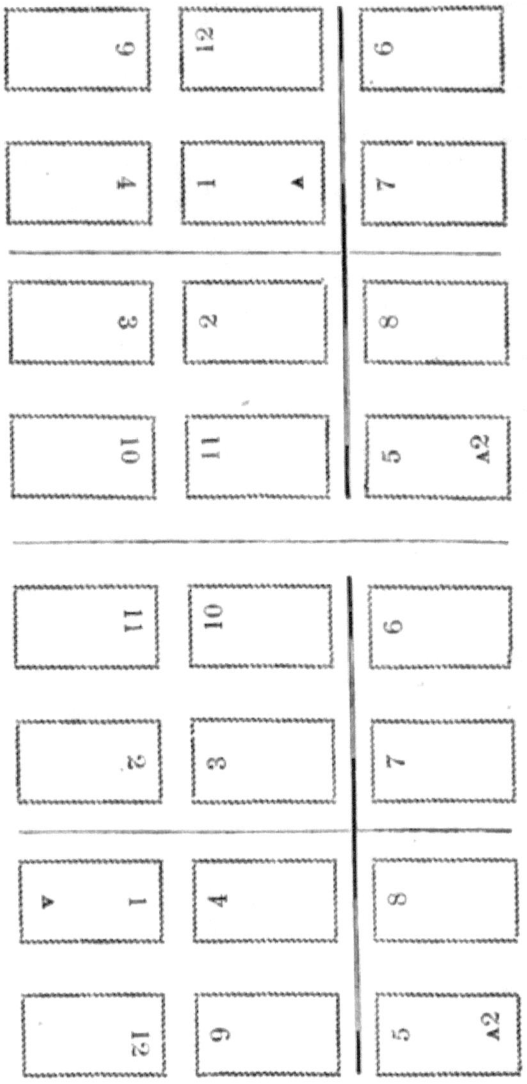

Different Methods of Imposing Half-Sheets of Twelves, from the Centre.

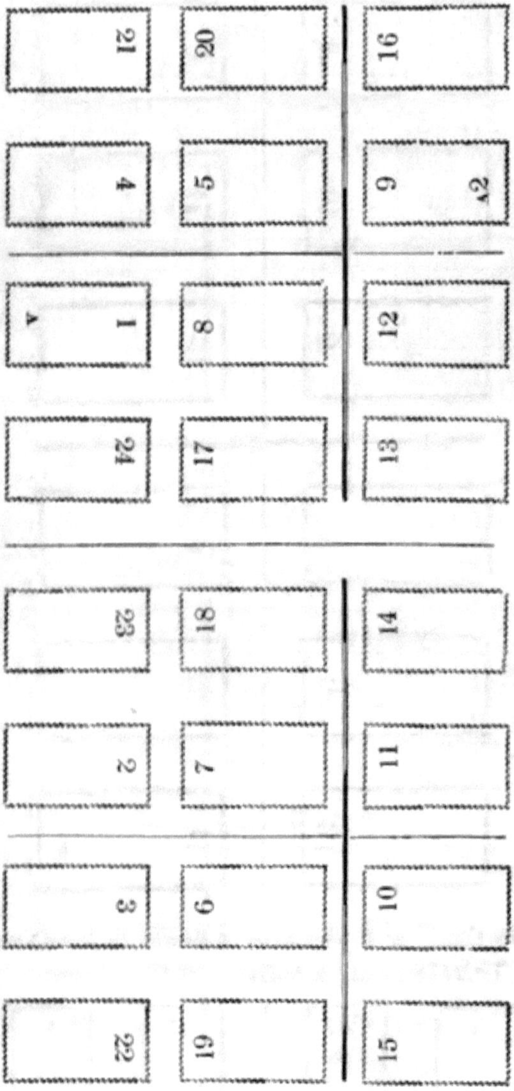

A Sheet of Twelves, Imposed from the Centre.

Outer Form of a Sheet of Long Twelves.

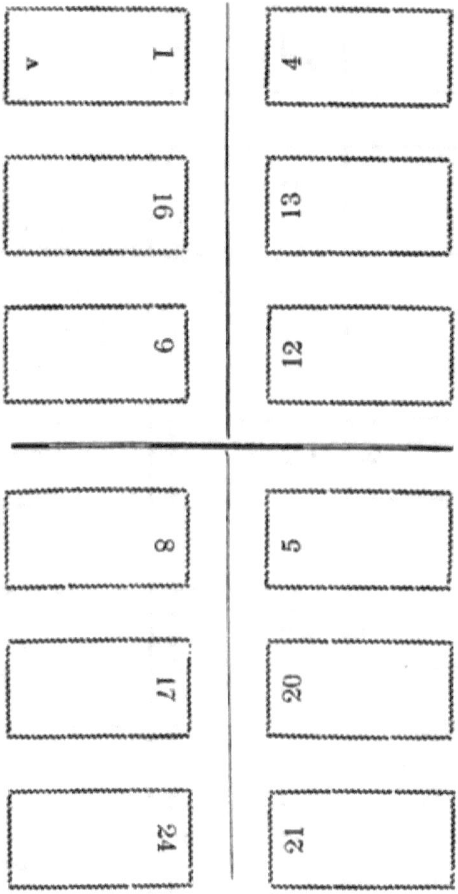

One-third, or 8 pages, of a Sheet of Twelves.

To be imposed as a slip, or in the off-cross.

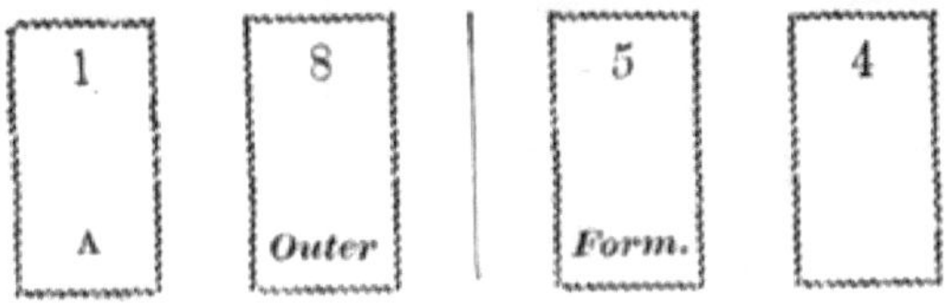

Inner Form of a Sheet of Long Twelves.

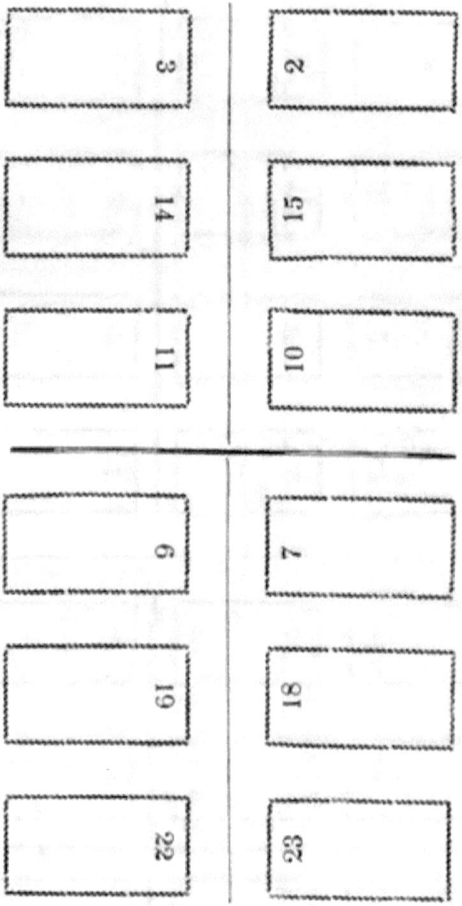

One-third, or 8 pages, of a Sheet of Twelves.

To be imposed as a slip, or in the off-cross.

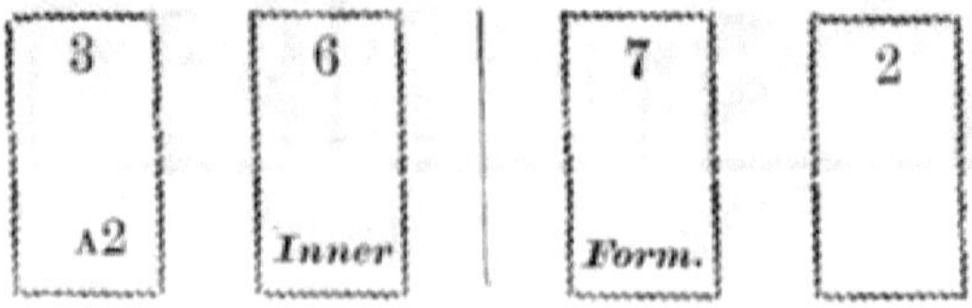

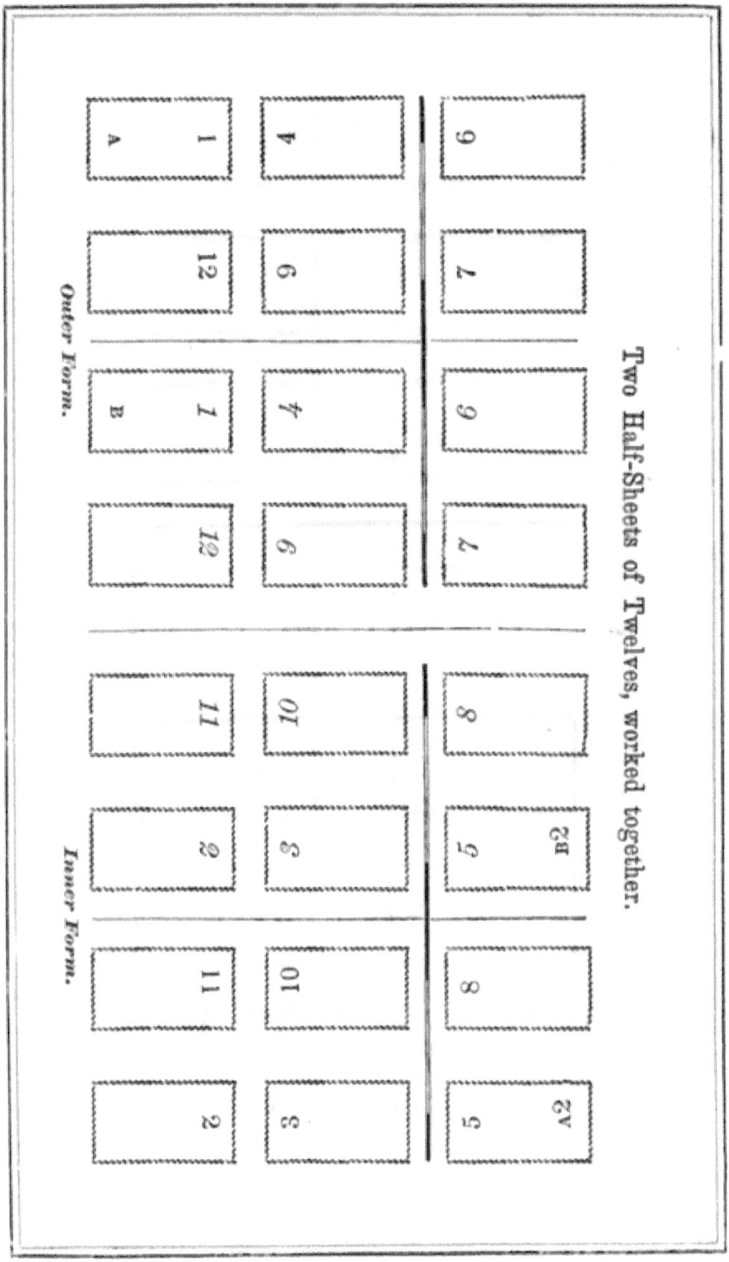

Two Half-Sheets of Twelves, worked together.

Half-Sheet of Twelves, with 2 Signatures.

4 pages of other matter.

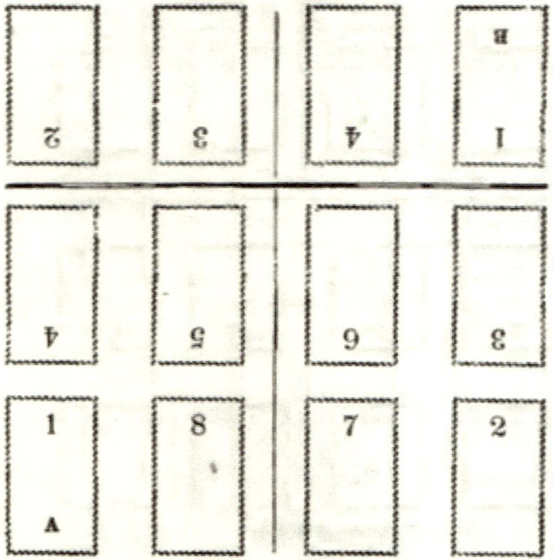

A Half-Sheet of Sixteens.

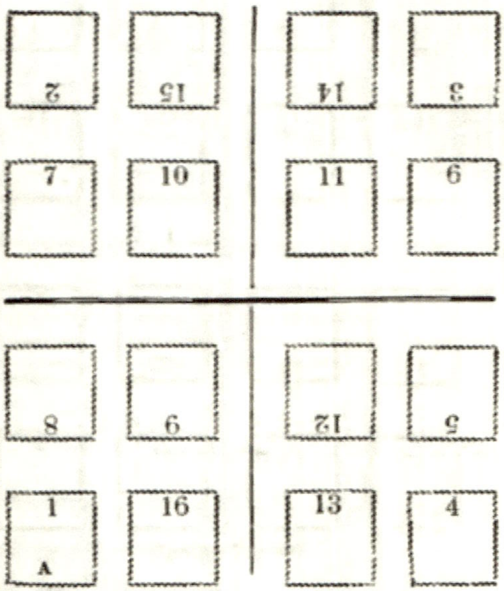

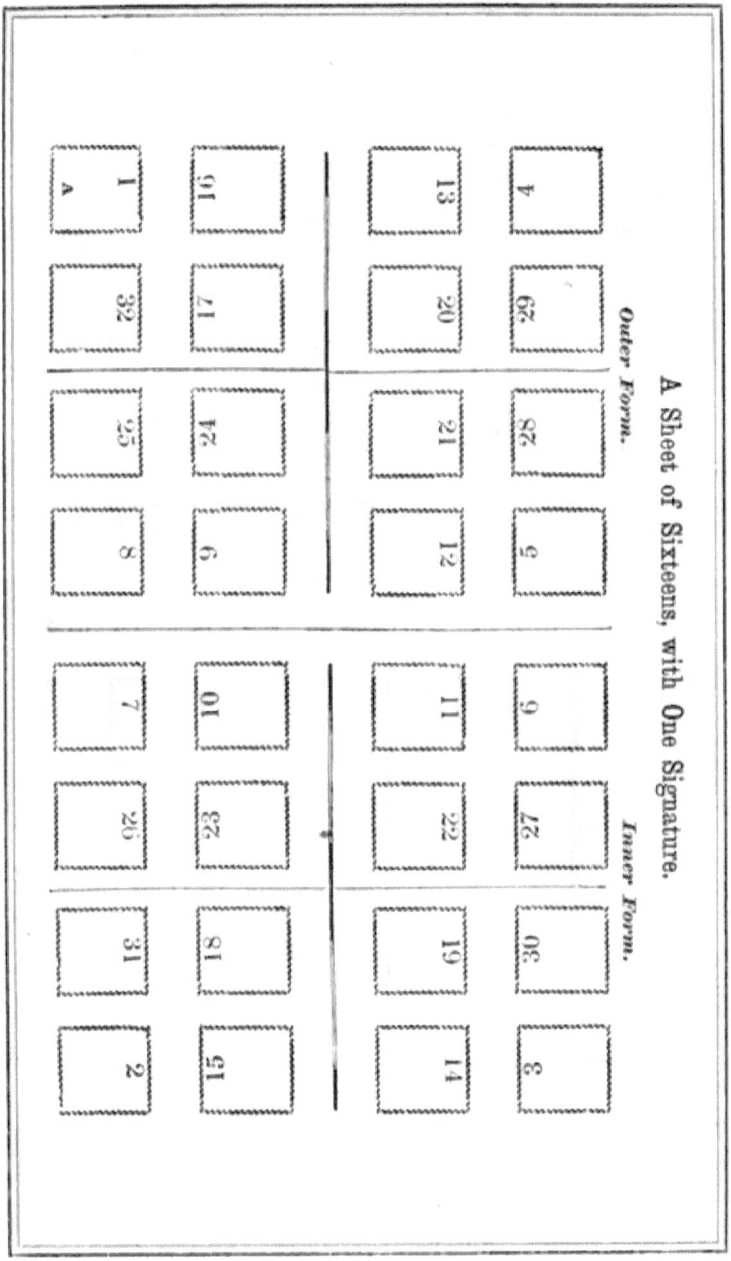
A Sheet of Sixteens, with One Signature.
Outer Form.
Inner Form.
A

A Half-Sheet of Eighteens.*

Containing 16 pages.

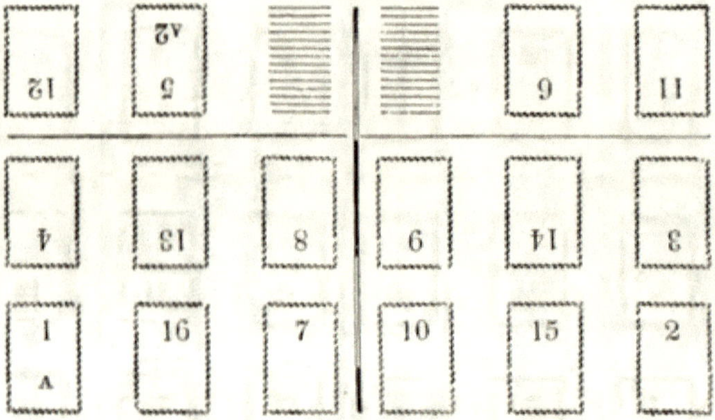

A Half-Sheet of Eighteens.†

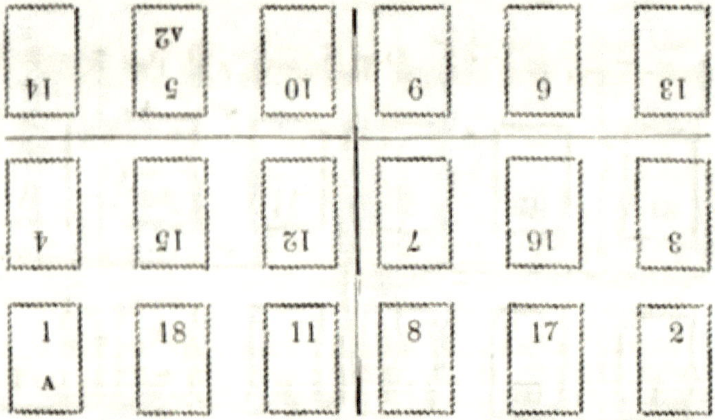

* The white paper of this half-sheet being worked off, the centre pages must be transposed,—viz. pages 7 and 10 in the room of 9 and 8, and pages 9 and 8 in the place of 7 and 10: when this is done, your imposition will be true.

† When the white páper is worked off, transpose the form,—viz. pages 11 and 8 in the room of 7 and 12, and pages 7 and 12 in the place of 11 and 8: this being done, the sheet will then fold up right.

Outer Form of a Sheet of Eighteens, to be folded together.

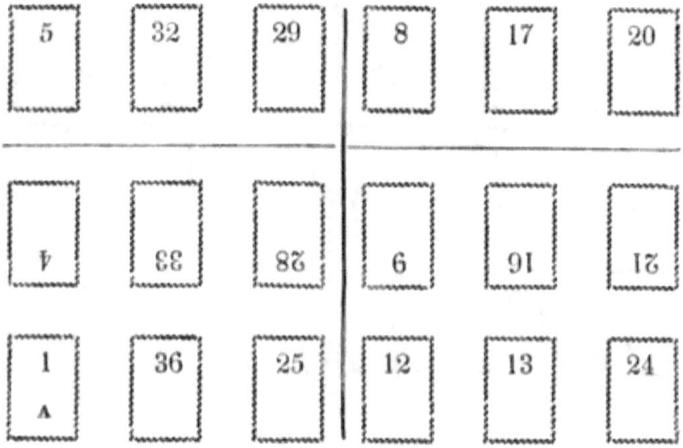

Outer Form of a Sheet of Eighteens, with One Signature.

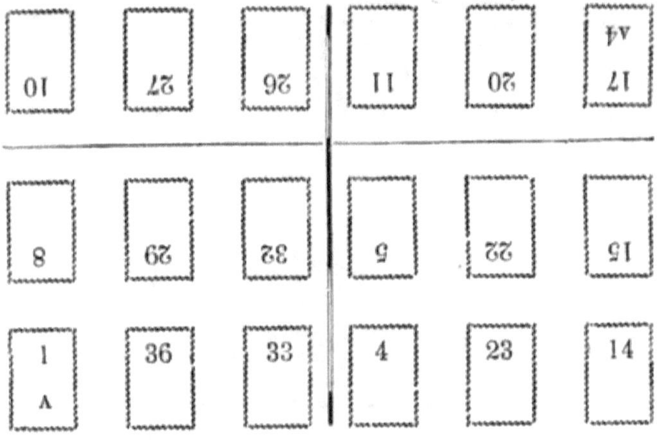

Inner Form of a Sheet of Eighteens, to be folded together.

19	18	7	30	31	6
22	15	10	27	34	3
23	14	11	26	35	2

Inner Form of a Sheet of Eighteens, with One Signature.

18	19	12	25	28	A2 9
16	21	6	31	30	7
13 A3	24	3	34	35	2

Outer Form of a Sheet of Eighteens, with Two Signatures.

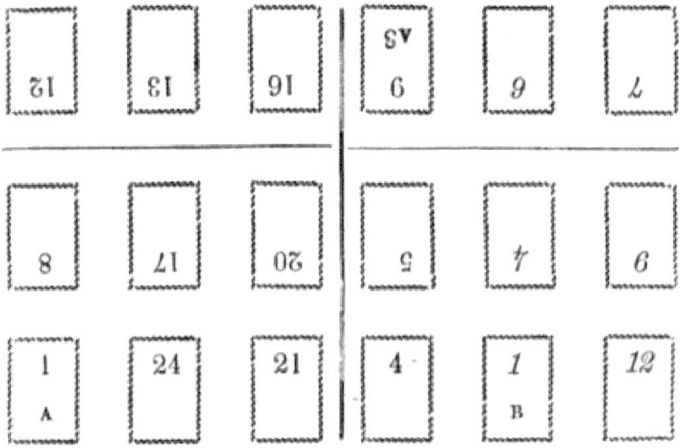

Outer Form of a Sheet of Eighteens, with Three Signatures.

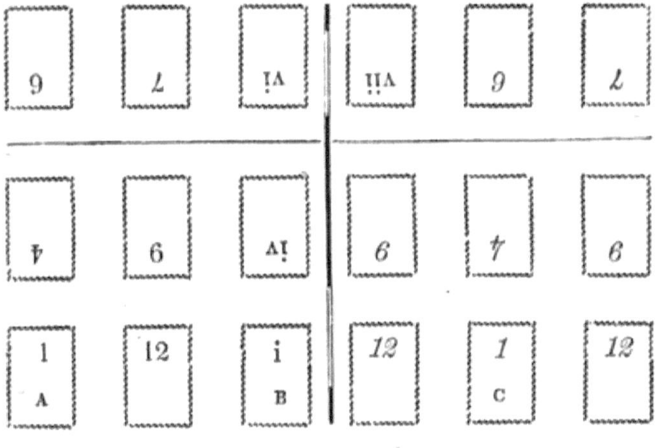

Inner Form of a Sheet of Eighteens, with Two Signatures.

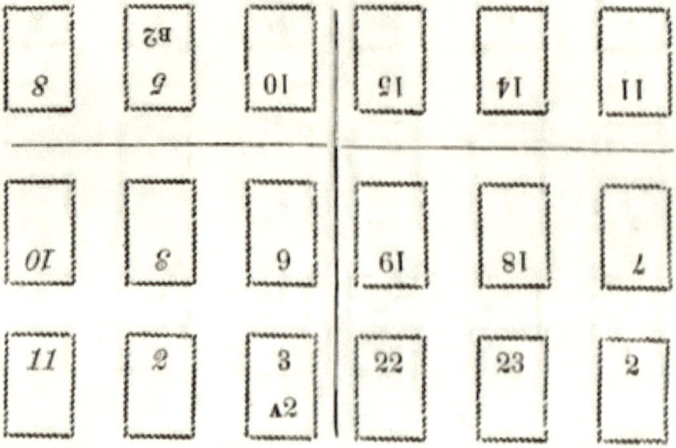

Inner Form of a Sheet of Eighteens, with Three Signatures.

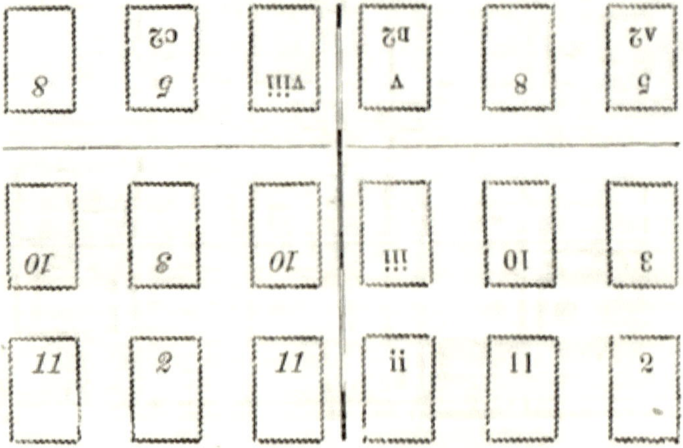

A Half-Sheet of Eighteens, without Transposition.*

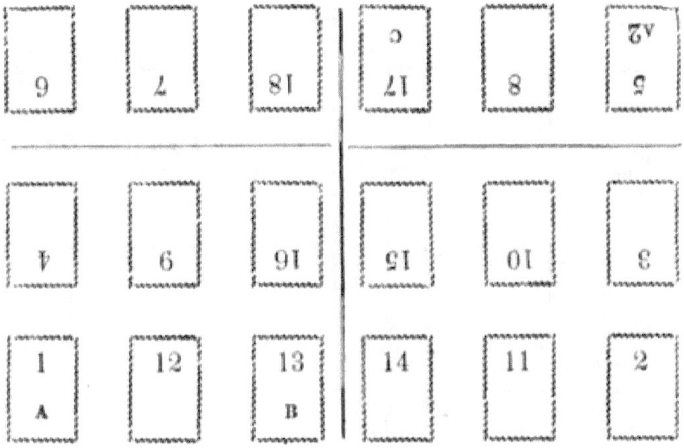

A Half-Sheet of Twenties, with Two Signatures.

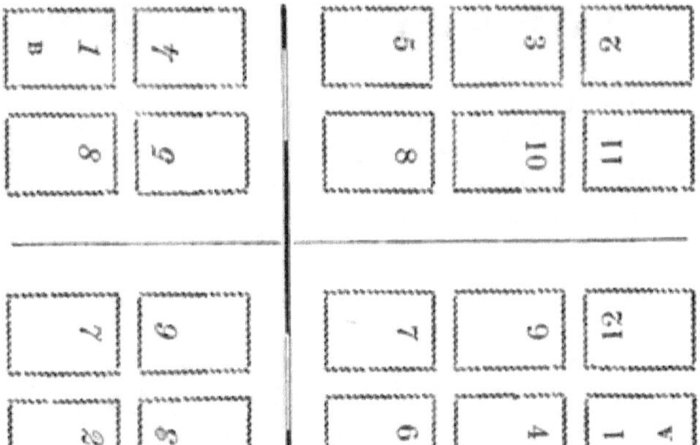

* This mode of imposition is very objectionable, as there will be, when the paper is cut up, three single leaves.

Inner Form of a Sheet of Twenties.

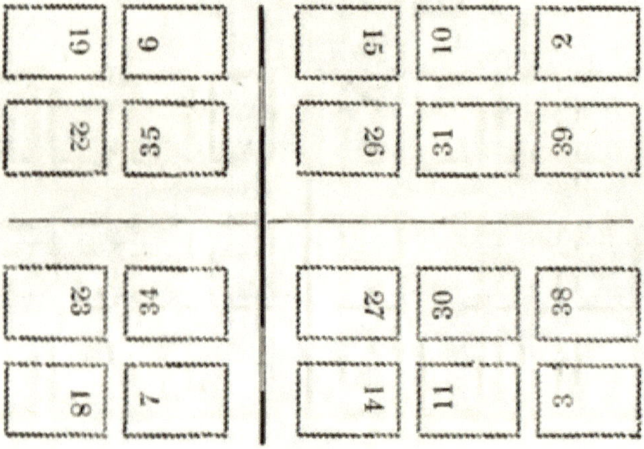

Outer Form of a Sheet of Twenties.

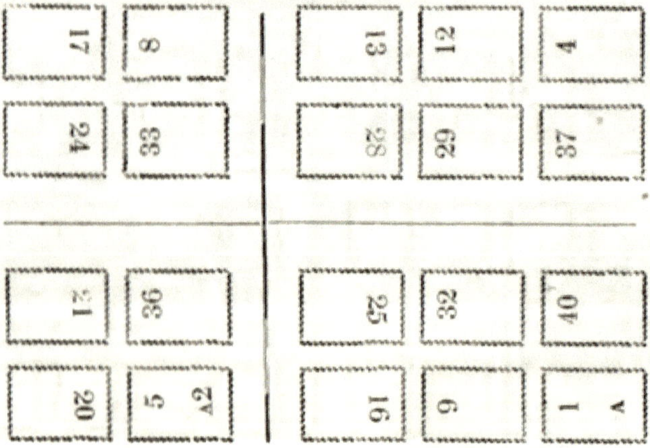

A Half-Sheet of Twenty-Fours.

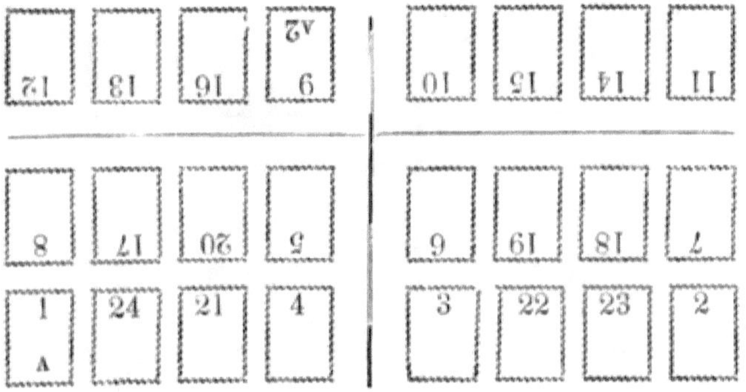

Outer Form of a Sheet of Twenty-Fours, with Two Signatures.

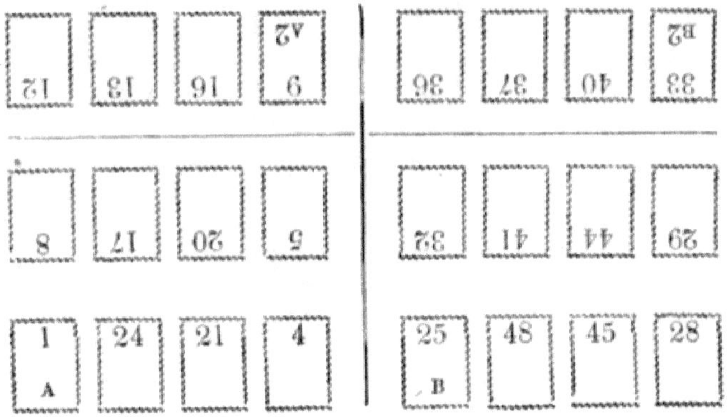

A Half-Sheet of Twenty-Fours, the Sixteen-way.

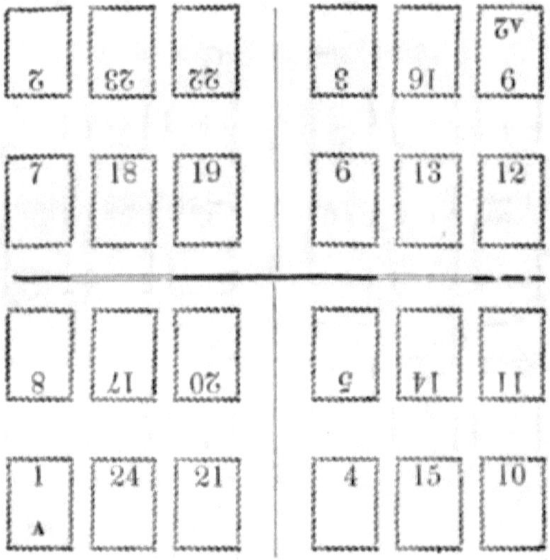

Inner Form of a Sheet of Twenty-Fours, with Two Signatures.

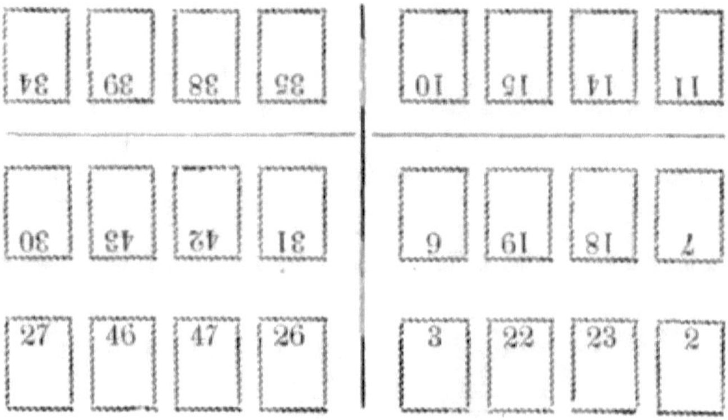

A Half-Sheet of Twenty-Fours, without Inset.

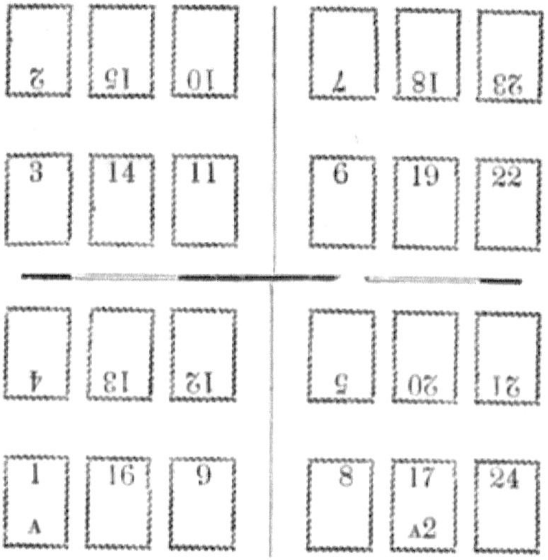

A Half-Sheet of Twenty-Fours, without Inset.

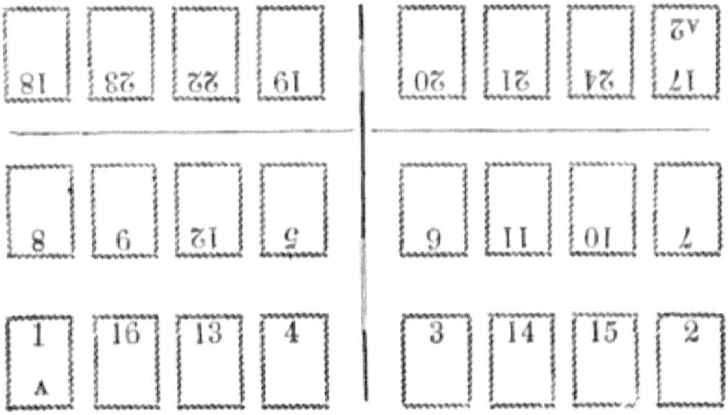

A Half-Sheet of Twenty-Fours, without Cutting.

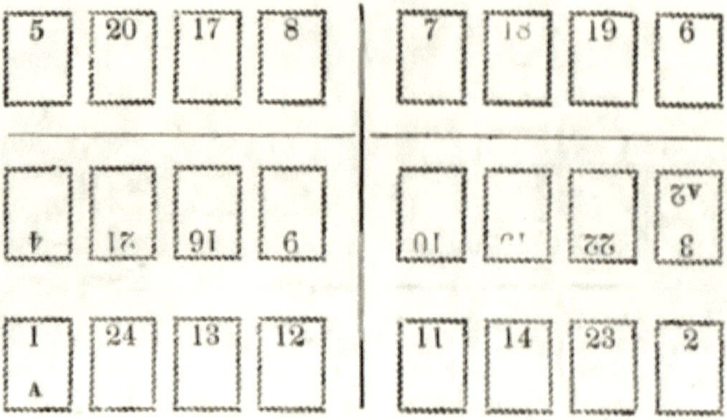

A Half-Sheet of Thirty-Twos.

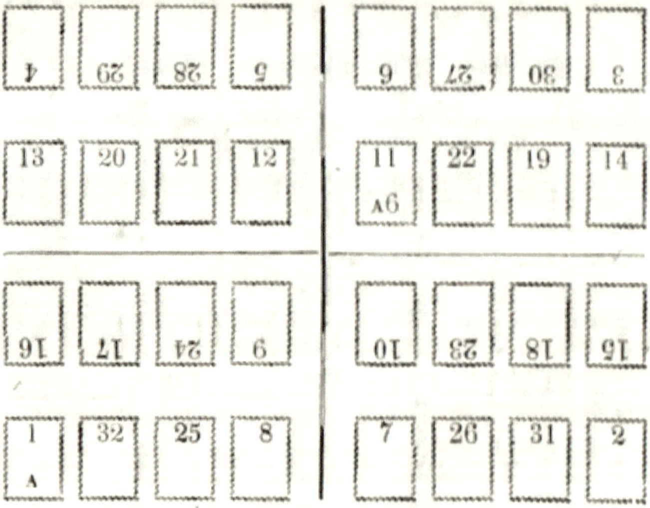

Outer Form of a Sheet of Thirty-Twos.

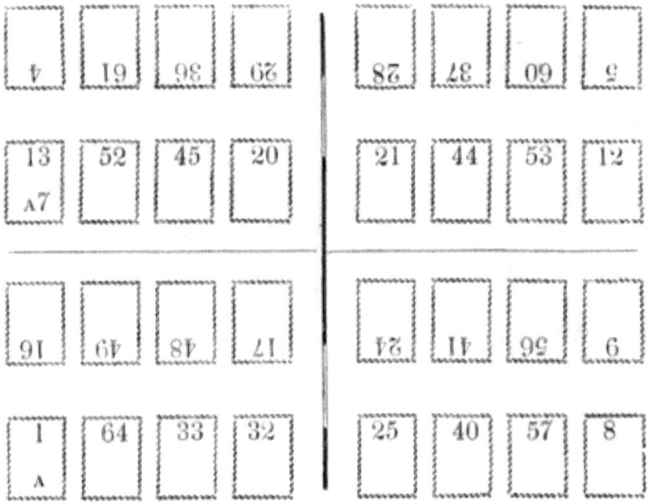

Outer Form of a Sheet of Thirty-Twos, with Four Sigs.

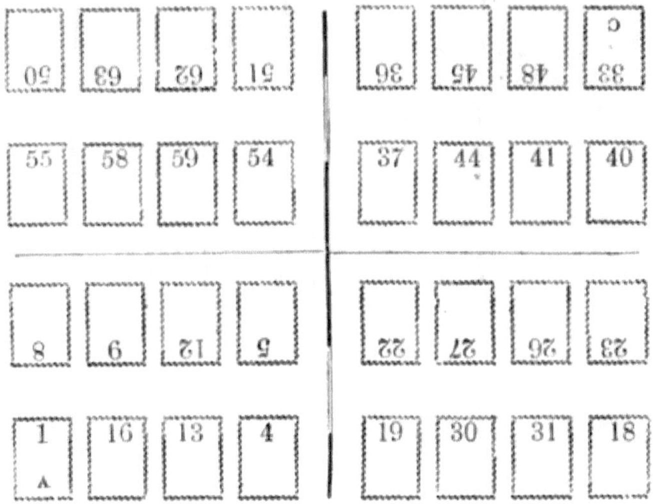

Inner Form of a Sheet of Thirty-Twos.

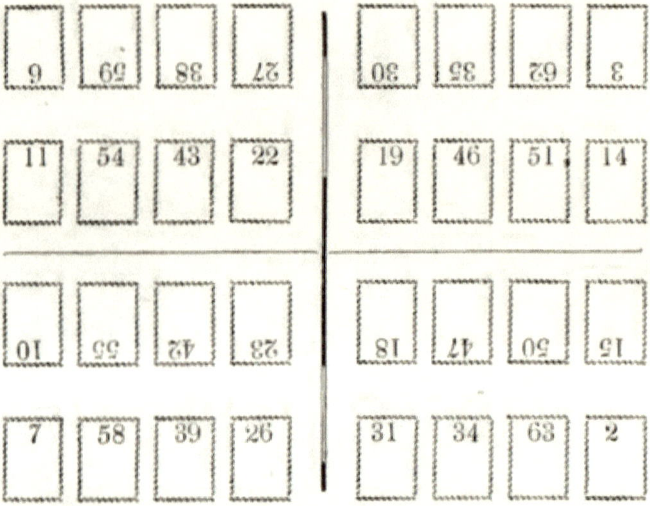

Inner Form of a Sheet of Thirty-Twos, with Four Sigs.

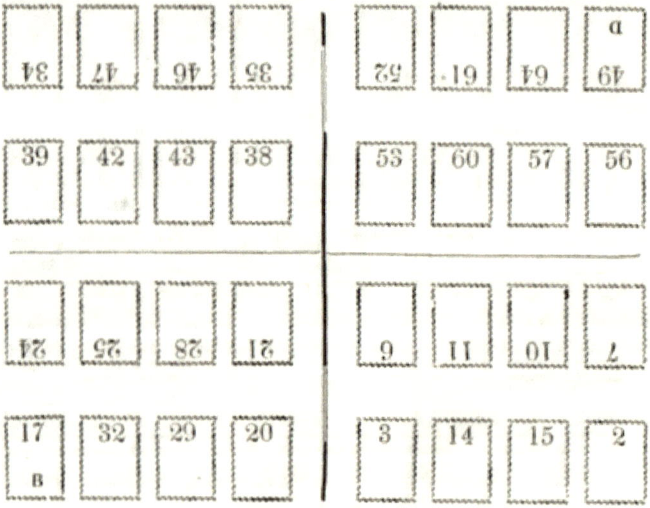

A Half-Sheet of Thirty-Twos, with Two Signatures.

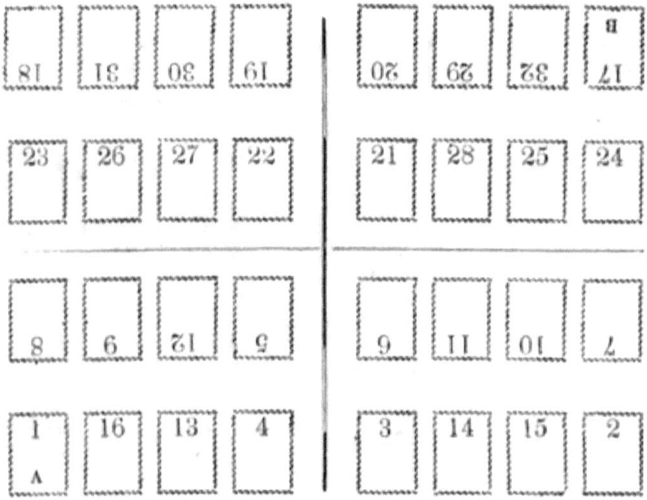

A Half-Sheet of Thirty-Twos, 20 pages of the Work, 4 pages of Title, &c., and 8 of other Matter.

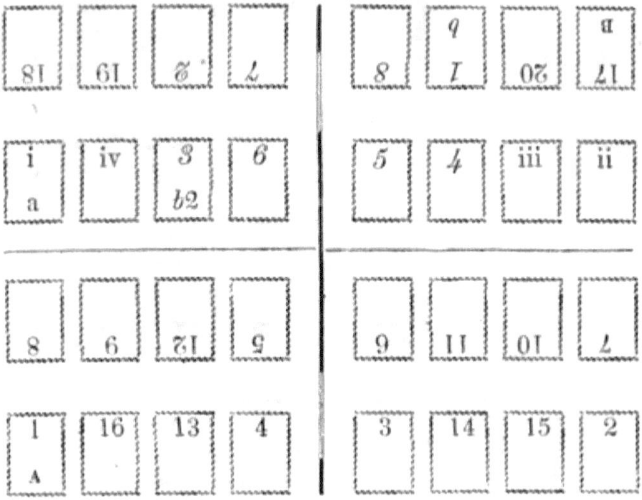

A Half-Sheet of Thirty-Sixes.

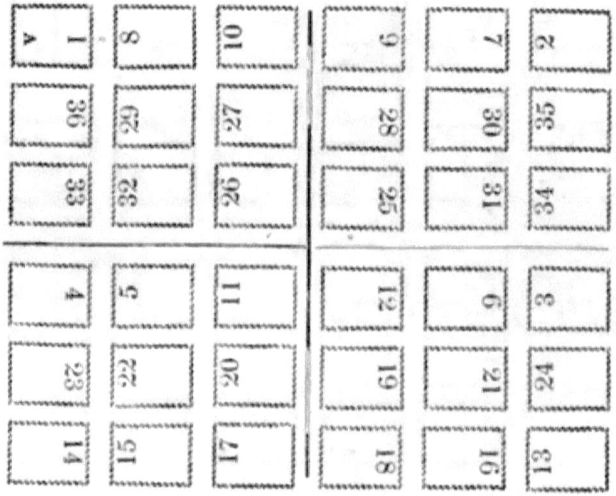

A Half-Sheet of Thirty-Sixes, without Cutting.

A Half-Sheet of Thirty-Sixes, with Two Signatures.

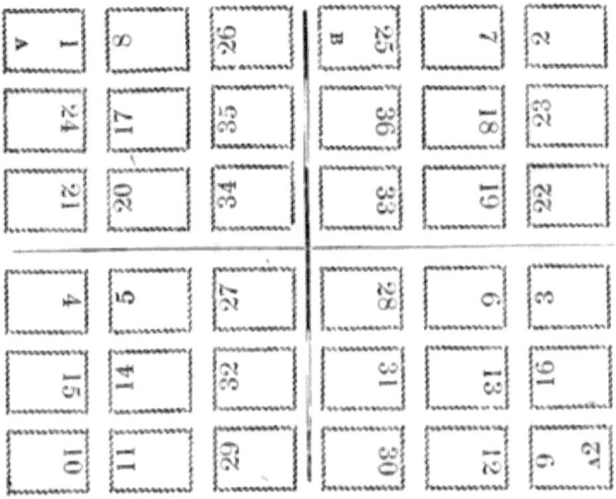

A Half-Sheet of Forties.

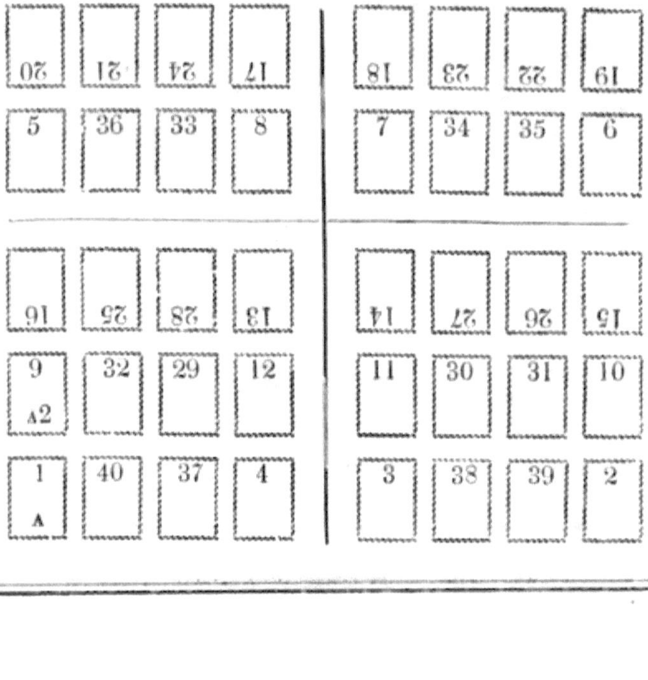

A Quarter-Sheet of Forty-Eights, with Two Signatures.

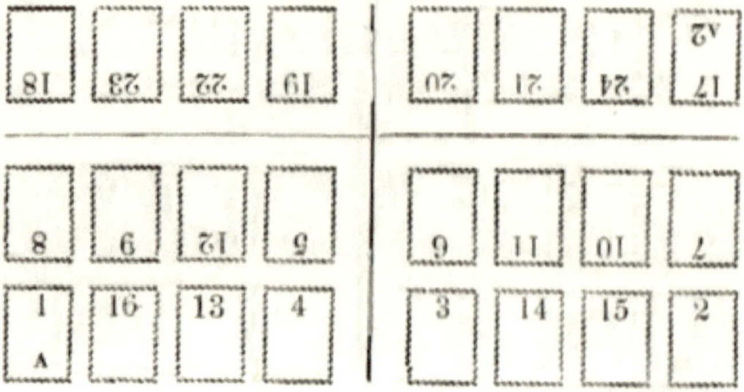

A Half Sheet of Forty-Eights, with Two Signatures.

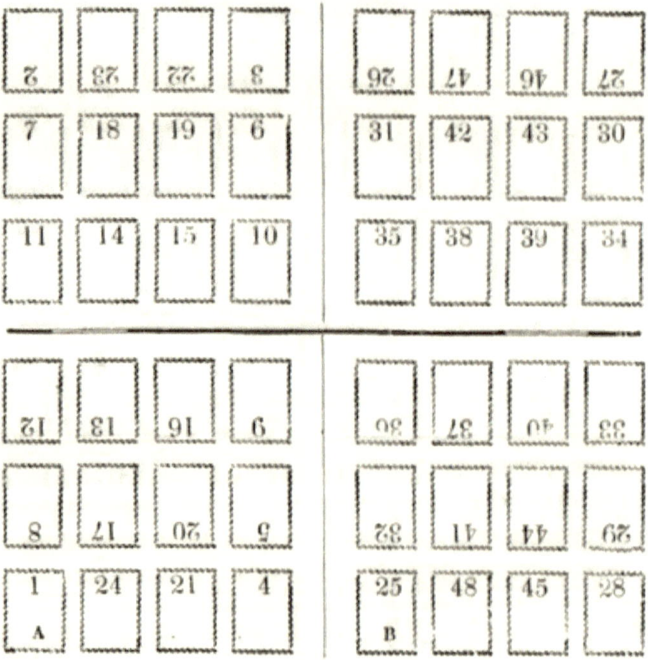

A Quarter-Sheet of Forty-Eights, without Cutting.

5	20	17	8	7	18	19	6
4	21	16	9	10	15	22	3
1 A	24	13	12	11	14	23	2

A Half-Sheet of Forty-Eights, with Three Signatures.

34	47	46	35	36	45	48	33 C
39	42	43	38	37	44	41	40
18	31	30	19	20	29	32	17 B
23	26	27	22	21	28	25	24
8	9	12	5	6	11	10	7
1 A	16	13	4	3	14	15	2

A Common Quarter-Sheet of Forty-Eights.

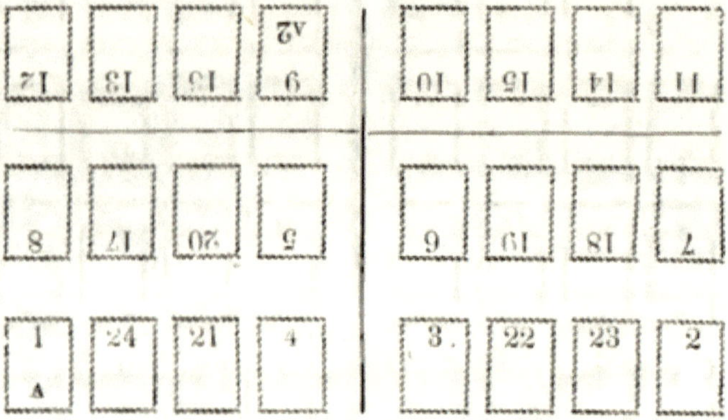

A Quarter-Sheet of Sixty-Fours, with Two Signatures.

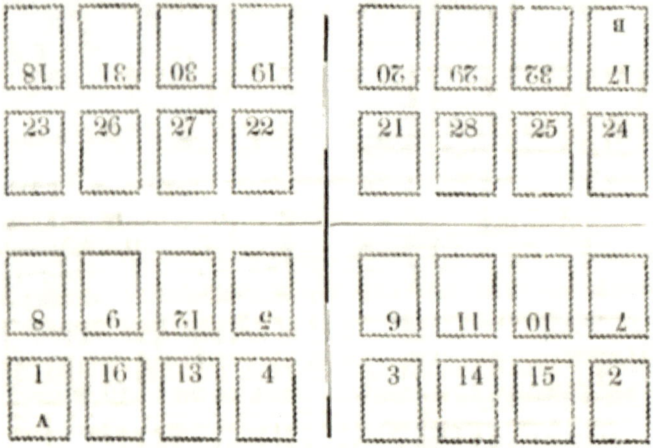

A Common Quarter-Sheet of Sixty-Fours.

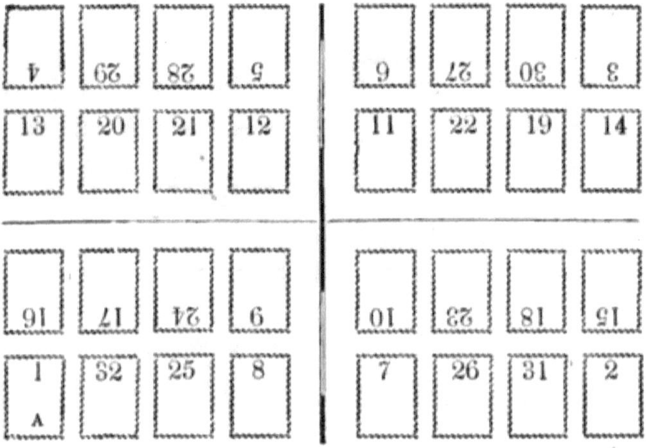

A Quarter-Sheet of Sixty-Fours, 20 pages of the Work, 8 of Title, and 4 of other Matter.

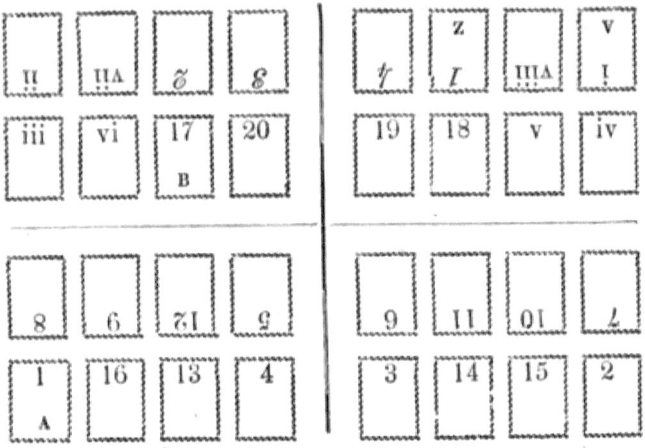

A Half-Sheet of Sixty-Fours.

2	63	34	31	26	39	58	7
15	50	47	18	23	42	55	10
14	51	46	19	22	43	54	11
3	62	35	30	27	38	59	6
4	61	36	29	28	37	60	5
13	52	45	20	21	44	53	12
16	49	48	17	24	41	56	9
1 A	64	33	32	25	40	57	8

A Half-Sheet of Seventy-Twos, with Three Signatures.

49 C	56	25 B	32	7	2
72	65	48	41	18	23
69	68	45	44	19	22
52	53	28	29	6	3
63	62	39	38	13	16
58	59	34	35	12	9
57	60	33	36	11	10
64	61	40	37	14	15
51	54	27	30	5	4
70	67	46	43	20	21
71	66	47	42	17	24
50	55	26	31	8	1 A

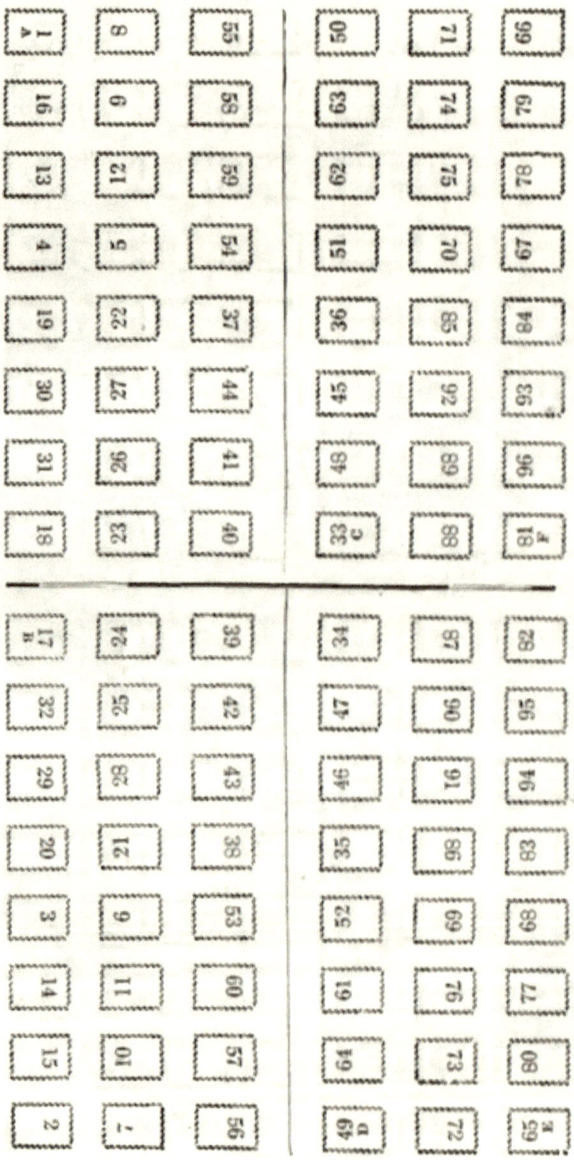

A Half-Sheet of Ninety-Sixes, with Six Signatures.

A Half-Sheet of One Hundred and Twenty-Eights, with Eight Signatures.

114	127	126	115	100	109	112	97 G	98	111	110	99	116	125	128	113 H
119	122	123	118	101	108	105	104	103	106	107	102	117	124	121	120
66	79	78	67	84	93	96	81 F	82	95	94	83	68	77	80	65 E
71	74	75	70	85	92	89	88	87	90	91	86	69	76	73	72
50	63	62	51	36	45	48	33 C	34	47	46	35	52	61	64	49 D
55	58	59	54	37	44	41	40	39	42	43	38	53	60	57	56
8	9	12	5	22	27	26	23	24	25	28	21	6	11	10	7
1 A	16	13	4	19	30	31	18	17 B	32	29	20	3	14	15	2

PROOF-READING AND CORRECTING.

PROOF-READING.

UNDENIABLE as is the fact that a book marred by typographical errors and grammatical blemishes is a scandal to the profession, it must be admitted that a careful, steady, and competent reader is indispensable in every printing-office.

It is eminently desirable that a reader should have been previously brought up a compositor. By a practical acquaintance with the mechanical departments of the business, he will be better able to detect those manifold errata which, unperceived by the man of mere learning and science, lie lurking, as it were, in a thousand different forms, in every sheet; and which, if overlooked, justly offend the taste and discernment of all appreciators of correct and beautiful typography.

Some of the principal imperfections which are more easily observed by the man of practical knowledge in the art of printing are the following: viz. imperfect, wrong-founted, and inverted letters, particularly the lower-case n, o, s, and the u, as well as p, d, b, and q; awkward and irregular spacing; uneven pages or columns; a false disposition of the reference marks; crookedness in words and lines; bad making-up of matter; erroneous indention, &c. These minutiæ, which are rather imperfections of workmanship than literal errors, are apt to be overlooked and neglected by mere literary readers.

Long and frequent habits of reading proof-sheets for the press, a quick eye, and a steady mind, will certainly enable a person, though not a compositor, to detect those minor deviations from correctness which the inexperienced and the careless are apt to overlook. But, while these habits are acquiring, without which no person can be safely intrusted to read a sheet for press, the labours of the printer are liable to go forth into the world in a manner that will reflect discredit on the employed and give offence to the employer. No form, therefore, ought to be put to press until it has been read and revised by an *experienced* reader.

A first-class proof-reader, in addition to a general and practical acquaintance with typography, should understand clearly the grammar and idiomatic structure of his mother-tongue, and have, as it were, an encyclopedic knowledge of the names, times, and productions of its writers, as well as a thorough familiarity with the Bible especially, and with Shakspeare. He should be, in fact, a living orthographical, biographical, bibliographical, geographical, historical, and scientific dictionary, with some smattering of Hebrew, Greek, Latin, French, Spanish, Italian, and German. Yet all these accomplishments are valueless unless he also possess a keen and quick eye, that, like a hound, can detect an error almost by scent. There are eyes of this sort, that with a cursory glance will catch a solitary error in a page. The world is little aware how greatly many authors are indebted to a competent proof-reader for not only reforming their spelling and punctuation, but for valuable suggestions in regard to style, language, and grammar,—thus rectifying faults which would have rendered their works fair game for the petulant critic.

Although no corrector of the press can strictly be required to do otherwise than to *follow his copy*,—that is, faithfully to adhere to the original, with all its defects,—yet every one must perceive that he performs a friendly, and perhaps a charitable, service, by pointing out, in proper time, imperfections and mistakes which have escaped the observation of a quick or voluminous writer. With the spirit, the opinions, the whims of an author, no corrector of the press has any business to interfere. In reprints of old and standard works, no license of alteration ought to be granted to either correctors or editors.

Strict uniformity should always be preserved in the use of capitals, in orthography, and punctuation. Nothing can be more vexatious to an author than to see the words *honour*, *favour*, &c. spelt with and without the *u*. This is a discrepancy which correctors ought studiously to avoid. The above observations equally apply to the use of capitals to noun-substantives, &c. in one place, and the omission of them in another. However the opinions of authors may differ in these respects, still the system of spelling, &c. must not be varied in the same work.

When an author gives him the option, a proof-reader ought to spell ambiguous words and arrange compounds in a methodical and uniform way; and, to enable the compositors to become acquainted with and to observe his method, he should furnish for their guidance a list of such ambiguous words and compounds.

Such being the qualifications of a reader, we exhibit the process which proof-sheets ought to undergo before the pages are put to press.

When a first proof is pulled, the compositor who imposed the sheet ought to collect and arrange the copy, and deliver both to the reader, who, after folding the sheet to prove the accuracy of its imposition, carefully examines the signatures, head-lines, and paging. He then calls his reading-boy, to read the copy aloud to him. This boy should be able to read with ease and distinctness any copy put into his hands. The eye of the reader should not follow, but rather precede, the voice of the boy; accustomed to this mode, he will be able to anticipate every single word in the copy; and, should a word or sentence happen to be missing in the proof, his attention will the more sensibly be arrested by it when he hears it pronounced by his reading-boy. He ought to be careful lest his eyes advance too far before the words of the boy; because, in his attention to the author's meaning, he will be apt to read words in the proof which do not actually appear there, and the accuracy of the reading-boy will but tend to confirm him in the mistake.

When the reading of the sheet is concluded, the number (if more than one) of the volume, signature, and *prima*, or first word of the ensuing sheet, should be accurately marked on the margin of the copy, and a bracket made before the first word of the next sheet, in order that the compositor, should he not have composed beyond the sheet, may know where to begin, without having the trouble of referring either to the proof or the form, and the reader will be certain that the commencement is right when he gets the succeeding sheet. This prevents unnecessary trouble both to the reader and compositor.

Before the proof is sent to the compositor to be corrected in the metal, an entry should be made in a book, according to the following plan:—

Date of reading.	Signatures.	Names of Works.	Sent out.	Returned.	Read for Press.
1865.			1865.	1865.	1865.
May 2	11	Decorative Printing...	May 2	May 4	May 5
" 4	82	Quarto Bible............	" 4	" 5	" 6
" 6	20	American Printer......	" 7	" 8	" 8
" 7	2	Life of Prescott........	" 8	" 9	" 9

This account being punctually kept, the reader can furnish the employer or overseer with an exact account of the state of each work, without delay or inconvenience.

After the compositors have corrected the errors in the form, a clean proof is pulled, which, with the first proof, is handed to the reader, who then collates the corrected sheet with the one before read, in order to ascertain whether the corrections have been properly made, and whether new errors have not been caused by negligence in the process; and, if the work be a reprint, or if the author is not to examine the proof, he then proceeds to read it very carefully for press.

Some proofs are so foul, that it is almost impossible for the compositor to correct all the marks at one time, and it is therefore necessary to have the neglected errors corrected and another sheet pulled before the proof is read finally. It not unfrequently happens that compositors, in the course of correcting, transpose a letter or word, or alter a letter in a word that is not marked, thus not only leaving one error uncorrected, but also making another; sometimes also, in respacing a line, a space is transposed or a hyphen is left in. Consequently it is absolutely necessary, in revising a proof, that the reader should not only look at the word marked, but he ought also to glance his eye over every line in which an alteration has been made.

In offices where two readers are employed, it is advisable that a proof-sheet should be read over by both; because the eye, in traversing the same ground, is liable to be drawn into mistake and oversight. The interest excited by the first reading having abated, a degree of listlessness imperceptibly steals upon the mind, which greatly endangers the correctness of a proof. Should *outs* or *doubles* occur in a proof, it ought to be again read by copy, to detect any improper correction in the overrunning or transposition of lines.

The duty of amending the punctuation should be generally confined to one reader. Where a compositor is liable, in this particular, to the whim or caprice of several readers, he certainly suffers injustice, because his time is unnecessarily frittered away; and not only is the work retarded, but the types are needlessly exposed to injury, to say nothing of the liability of creating fresh errors, &c.

Before a manuscript is brought to the printer, it ought to be as perfect as the author can make it. The compositor is bound to "follow the copy," in word and sentiment, unless, indeed, he meets with instances of wrong punctuation or false grammar (and such instances are not rare), which his intelligence enables him to amend. After the matter has been read and corrected in the office, a proof is sent to the author; and, if it corresponds with the copy, the compositor's responsibility is at an end. He has done all he is paid for; and, should the author desire any changes made in his matter, of course he must pay for them.

Sentiments in print look marvellously different from the same ideas in manuscript; and we are not surprised that writers should wish to polish a little; nor do we object to their natural desire of amending or beautifying their mental products. But let them not forget that pay-time will come,—when the item for alterations will loom out with a startling distinctness in their bill. They found it easy in the proof to erase a word or two here and insert a word or two there; without dreaming, perhaps, that in consequence of these little erasures and insertions the compositor would be compelled to alter and reconstruct much of his work. We know of a volume on which the alterations alone have consumed time equal to one man's work for nearly two and a half years. How unreasonable—nay, how transparently unjust—the expectation that the printer should give gratuitously the time and trouble requisite for the radical changes in the type which an author's whim or taste may demand!

Stower says, "It may not be improper, in this place, just to take notice of the great danger to the correctness of a work which arises from the practice, too common with some authors, of keeping their proof-sheets too long in their hands before they are returned to the printer. As the pages in the metal get dry, the adhesion of the types to each other is weakened, and the swell or extension of the quoins and furniture, which the moisture had occasioned, is removed; so that there is great danger of letters falling out when a form is long kept from the press. Nor is the danger which is hereby occasioned to correctness the only inconvenience: the impatience of authors to see their works in a fit state for publication is almost proverbial. The pleasure arising from beholding, as it were, the 'form and texture' of one's thoughts, is a sensation much easier felt than described. That authors, therefore, may partake of this pleasure in a speedy and regular succession, they should make a point of forwarding their proof-sheets to the printer as quick as possible, not only that they may the sooner be got ready for the press, but that the work may proceed in a regular manner, without being interrupted by the forwarding of other works in lieu of that the proof-sheets of which are detained beyond the proper time in the hands of the author.

"Authors are very apt to make alterations, and to correct and amend the style or arguments of their works, when they first see them in print. This is certainly the worst time for this labour, as it is necessarily attended with an expense which, in large works, will imperceptibly swell to a serious sum; when, however, this method of alteration is adopted by an author, the reader must always be careful to read the whole sheet over once more with very great attention before it is finally put to press.

"A proof-sheet having duly undergone this routine of purgation, may be supposed to be as free from errata as the nature of the thing will admit, and the word 'Press' may be written at the top of the first page of it. This is an important word to every reader: if he have suffered his attention to be drawn aside from the nature of his proper business, and errors should be discovered when it is too late to have them corrected, this word 'Press' is as the signature of the death-warrant of his reputation. A reader, therefore, should be a man of one business,—always upon the alert,—all eye,—all attention. Possessing a becoming reliance on his own powers, he should never be too confident of success. Imperfection clings to him on every side. Errors and mistakes assail him from every quarter. His business is of a nature that may render him obnoxious to blame, but can hardly be said to bring him in any very large stock of praise. If errors escape him, he is justly to be censured; for perfection is his duty. If his labours are wholly free from mistake,—which is, alas! a very rare case,—he has done no more than he ought, and, consequently, can merit only a comparative degree of commendation, in that he had the good fortune to be more successful in his labours after perfection than some of his brethren in the same employment."

The form being finally laid on the press, and a revise pulled by the pressman, he sends it to the overseer, who carefully examines whether all the marks have been attended to, and looks along the sides and heads of the respective pages, to observe whether any letter has fallen out, any crookedness in the locking up of the form, any battered letters, or any *bite* from the frisket. Should the revise prove faultless, he returns it to the pressman, with the word "Revise" written on the margin; if otherwise, to the compositor to whom the form belongs, for immediate correction.

TYPOGRAPHICAL MARKS EXEMPLIFIED.

THE following table of proof-marks will be appreciated by authors. Due attention to the explanations will insure an apt proficiency in the manual department of proof-reading.

1 a/ THOUGH a veriety of opinions exist as to the individual by wyom the art of printing was 2 G.
first discovered; yet all authorities concur in
admitting Peter Schoeffer to be the person 3 Caps
who invented *cast metal types*, having learned
the art of of *cutting* the letters from the Gut-
5 tembergs he is also supposed to have been
6 the first whoengraved on copper plates. The 7 /-/
following testimony is preseved in the family, 8 r/
9 by Jo. Fred. Faustus, of Ascheffenburg:
10 'Peter Schoeffer, of Gernsheim, perceiving <# 6
11 his master Fausts design, and being himself S. Caps.
12 tr. (desirous ardently) to improve the art, found
out (by the good providence of God) the
method of cutting (*incidendi*) the characters 13 stet.
in a *matrix*, that the letters might easily be
5 singly *cast* instead of bieng *cut*. He pri- 12 ei/
14 vately *cut matrices* for the whole alphabet: 15
Faust was so pleased with the contrivance,
that he promised Peter to give him his only 17 wf.
16 daughter Christina in marriage a promise 3 Ital.
which he soon after performed.
18 no ¶
19 as/ (But there were many difficulties at first
with these *letters*, as there had been before 3 Rom.
20 + with wooden ones, the metal being too soft 3 Ital.
to support the force of the im pression: but 9
this defect was soon remedied, by mixing
a substance with the metal which sufficiently tr. 12
5 hardened it

and when he showed his master the letters cast from these matrices,

THOUGH a variety of opinions exist as to the individual by whom the art of printing was first discovered; yet all authorities concur in admitting PETER SCHOEFFER to be the person who invented *cast metal types*, having learned the art of *cutting* the letters from the Guttembergs: he is also supposed to have been the first who engraved on copper-plates. The following testimony is preserved in the family, by Jo. Fred. Faustus, of Ascheffenburg:

'PETER SCHOEFFER, of Gernsheim, perceiving his master Faust's design, and being himself ardently desirous to improve the art, found out (by the good providence of God) the method of cutting (*incidendi*) the characters in a *matrix*, that the letters might easily be singly *cast*, instead of being *cut*. He privately *cut matrices* for the whole alphabet: and when he showed his master the letters cast from these matrices, Faust was so pleased with the contrivance, that he promised Peter to give him his only daughter *Christina* in marriage, a promise which he soon after performed. But there were as many difficulties at first with these letters, as there had been before with *wooden ones*, the metal being too soft to support the force of the impression: but this defect was soon remedied, by mixing the metal with a substance which sufficiently hardened it.'

EXPLANATION OF THE CORRECTIONS.

A wrong letter in a word is noted by drawing a short perpendicular line through it, and making another short line in the margin, behind which the right letter is placed. (See No. 1.) In this manner whole words are corrected, by drawing a line across the wrong word and making the right one in the margin opposite.

A turned letter is noted by drawing a line through it, and writing the mark No. 2 in the margin.

If letters or words require to be altered from one character to another, a parallel line or lines must be made underneath the word or letter,—viz. for capitals, three lines; small capitals, two lines; and Italic, one line; and, in the margin opposite the line where the alteration occurs, *Caps*, *Small Caps*, or *Ital.* must be written. (See No. 3.)

When letters or words are set double, or are required to be taken out, a line is drawn through the superfluous word or letter, and the mark No. 4 placed opposite in the margin.

Where the punctuation requires to be altered, the correct point, marked in the margin, should be encircled. (See No. 5.)

When a space is omitted between two words or letters which should be separated, a caret must be made where the separation ought to be, and the sign No. 6 placed opposite in the margin.

No. 7 describes the manner in which the hyphen and ellipsis line are marked.

When a letter has been omitted, a caret is put at the place of omission, and the letter marked as No. 8.

Where letters that should be joined are separated, or where a line is too widely spaced, the mark No. 9 must be placed under them, and the correction denoted by the marks in the margin.

Where a new paragraph is required, a quadrangle is drawn in the margin, and a caret placed at the beginning of the sentence. (See No. 10.)

No. 11 shows the way in which the apostrophe, inverted commas, the star and other references, and superior letters and figures, are marked.

Where two words are transposed, a line is drawn over one word and below the other, and the mark No. 12 placed in the margin; but where several words require to be transposed, their right order is signified by a figure placed over each word, and the mark No. 12 in the margin.

Where words have been struck out that have afterward been approved of, dots should be marked under them, and *Stet* written in the margin. (See No. 13.)

Where a space sticks up between two words, a horizontal line is drawn under it, and the mark No. 14 placed opposite, in the margin.

Where several words have been left out, they are transcribed at the bottom of the page, and a line drawn from the place of omission to the written words (see No. 15); but if the omitted matter is too extensive to be copied at the foot of the page, *Out, see copy*, is written in the margin, and the missing lines are enclosed between brackets, and the word *Out* is inserted in the margin of the copy.

Where letters stand crooked, they are noted by a line (see No. 16); but, where a page hangs, lines are drawn across the entire part affected.

When a smaller or larger letter, of a different fount, is improperly introduced into the page, it is noted by the mark No. 17, which signifies wrong fount.

If a paragraph is improperly made, a line is drawn from the broken-off matter to the next paragraph, and *No* ¶ written in the margin. (See No. 18.)

Where a word has been left out or is to be added, a caret must be made in the place where it should come in, and the word written in the margin. (See No. 19.)

Where a faulty letter appears, it is marked by making a cross under it, and placing a similar one in the margin (see No. 20); though some prefer to draw a perpendicular line through it, as in the case of a wrong letter.

CORRECTING IN THE METAL.

Correcting is the most disagreeable part of a compositor's business, diminishing as it does his earnings, and causing great fatigue, and, by leaning over the stone, prejudicing his health. A foul proof, however, is a fault without extenuation, and seems to deserve some punishment. The noise and confusion which prevail in badly governed printing-offices, from light and frivolous conversation, not only retard business, but distract the attention of the compositor from the subject he has in hand, and cause him to make many mistakes. Some men, no doubt, can support a conversation and at the same time compose correctly; but their noise confuses those who are unable to preserve accuracy except by close attention to their copy in silence.

The first proof should contain merely the errors of the compositor; but it frequently happens that the corrector heightens them by his peculiarities. When this is unnecessarily done, it is an act of injustice to the compositor: it is sufficient for him to rectify such mistakes as arise either from inattention to his copy or want of judgment. The compositor ought not to suffer from the humour of a reader in capriciously altering commas and semicolons in the first proof (unless the sense is perverted), which he not unfrequently re-alters in the second, from a doubt as to the propriety of the points to be adopted.

When a proof is handed to the compositor, he should immediately correct it; and the reader, correlatively, should be equally prompt in his department. Can it reasonably be expected that the compositor will feel inclined to forward his proof, when he knows that the reader will delay it for hours?

Should a compositor have transposed two or more pages, either from an error in the folios or any other cause, he must unlock the quarter containing them, and, loosening the cross or crosses from the furniture, he lifts the chase and the remaining quarters off the stone. Should he have furniture sufficient round each page, he may move them into their proper stations by pressing the balls of his thumbs and fingers against the furniture at the head, foot, and sides of each page. If the letter be small, it will be advisable to wet the pages, because few imposing-stones are horizontal, or so steady that they will not shake when touched, or by the motion of the floor, occasioned by persons walking or dragging forms.

Should a compositor find that his pages *hang*, he must unlock the quarter, and pat the face of the type with the balls of his fingers until he gets it into a square position.

When a compositor unlocks a form, he should be careful not to leave the unlocked quoins too slack, as the force necessary to loosen the others may *squabble* the matter, or occasion it to *hang*.

A compositor should possess the following requisites before he begins to correct:—

"What is required of a compositor when he goes about correcting a foul proof, is a sharp bodkin and patience; because, without them, the letter cannot escape suffering by the steel, and hurrying will not permit him to justify the lines true. No wonder, therefore, to see pigeon-holes in one place, and pi in another."*

When the compositor has as many corrections between the thumb and forefinger of his left hand as he can conveniently hold, or, what is better, in his composing-stick (beginning at the bottom of the page, in order that they may follow regularly), and an assortment of spaces on a piece of paper, or, what is more convenient, in a small square box with partitions in it,—let him take the bodkin in his right hand, and, instead of raising each letter he may have to alter, place the point of the bodkin at one end of the line, and, with the forefinger of his left hand against the other, raise the whole line sufficiently high to afford him a clear view of the spacing; he may then change the faulty letter and alter his spacing before he drops the line. By this method he will not injure the type, which he must do if he force the bodkin into their sides or heads; a greater degree of regularity is insured where there may be occasion to alter the spacing, and no more time is taken up than by the other method.

In tables, and other matter, where rules prevent the lines from being raised, the letters must be drawn up by the bodkin; this is done by the compositor holding the instrument fast in his right hand, with the blade between his forefinger and thumb within about three-quarters of an inch from the point: thus guiding it steadily to the faulty letter, he sticks the point of the bodkin into the neck of the letter between the beard and the face, and draws it up above the other types, so that he can take it out with the forefinger and thumb of his left hand. In performing this operation, the blade of the bodkin should

---

* The following epitaph was no doubt written by a printer while performing the most disagreeable task attendant on his profession:—

No more shall copy bad perplex my brain,
No more shall type's small face my eyeballs strain;
No more the proof's foul page create me troubles,
By errors, transpositions, outs, and doubles:
No more my head shall ache from authors' whims,
As overrunnings, driving-outs, and ins;
The sturdy pressman's frown I now may scoff,
Revised, corrected, finally wrought off.

be kept as flat as possible on the face of the type, but it should not touch any of the surrounding types, as the slightest graze imaginable will injure their face, and they will consequently appear imperfect in the next proof, when he will have the trouble of altering them, his employer suffering the loss of the type.

The bodkin blade being held almost flat to the form, a small horizontal entrance of its point into the neck of the letter will raise it above the face of the form; but, if the bodkin be held nearly upright, it will not have sufficient purchase to draw the letter up, because the weight of the type and its close confinement will have greater power than the sharp point of the steel. By pressing sidewise, the bodkin blade acts as a lever, even though it has no other purchase than the slight motion of the hand.

The most careful compositor cannot at all times avoid leaving a word out, or composing the same word twice. When this happens, he should consider the best mode of rectifying the accident, by driving out or getting in, either above the error or below it. This ascertained, let the matter be taken upon a galley, and overrun in the composing-stick. Overrunning on the stone is an unsafe, unworkmanlike, and dilatory method, destroying the justification, and rendering the spacing uneven.

In correcting, care should be taken to avoid hair-spacing a line, by overrunning either back or forward. In overrunning the matter, the division should be used as little as possible; for, though the compositor may carefully follow the instructions laid down in this work on the subject of spacing and dividing, yet the effect of his attention will be completely destroyed if not followed up at the stone.

We here emphatically remark that, if authors were careful to spell properly the names of persons and places, technical and scientific terms, &c., and to write legibly, marking the end of sentences clearly, the work of the compositor would be facilitated, many errors would be prevented, and time, temper, and expense greatly economized.

Further. Let us remind authors that every correction made on their proof that is a variation from the copy as furnished to the printer is charged for according to the time required to make it. The justice of the charge is obvious; yet, strange to say, there is probably no item so frequently disputed by publishers. A man employs a mechanic to build a house according to fixed specifications; but, in the course of its erection, he improves or changes the plan, and orders certain portions to be torn down and rebuilt: is the mechanic to bear the loss? Certainly not. So, when a compositor builds up his page of type according to the copy furnished, he is right in requiring compensation for alterations made in it. He is not to suffer for the author's desire to improve his intellectual edifice.

TABLE OF SIGNATURES.

On the two following pages will be found a complete list of signatures for books in octavo, twelves and eighteens, sixteens, and twenty-fours.

The 24mo signatures in this table are arranged to bring the second signature on either the 9th or 17th page of the form. If the sheet is to be folded as an 8vo and 16mo, the figure signatures may be used; but if as two 12mos, the letter signatures will be used.

17

8vo.

1	1	A	481	61	3 L
9	2	B	489	62	3 M
17	3	C	497	63	3 N
25	4	D	505	64	3 O
33	5	E	513	65	3 P
41	6	F	521	66	3 Q
49	7	G	529	67	3 R
57	8	H	537	68	3 S
65	9	I	545	69	3 T
73	10	K	553	70	3 U
81	11	L	561	71	3 V
89	12	M	569	72	3 W
97	13	N	577	73	3 X
105	14	O	585	74	3 Y
113	15	P	593	75	3 Z
121	16	Q	601	76	4 A
129	17	R	609	77	4 B
137	18	S	617	78	4 C
145	19	T	625	79	4 D
153	20	U	633	80	4 E
161	21	V	641	81	4 F
169	22	W	649	82	4 G
177	23	X	657	83	4 H
185	24	Y	665	84	4 I
193	25	Z	673	85	4 K
201	26	2 A	681	86	4 L
209	27	2 B	689	87	4 M
217	28	2 C	697	88	4 N
225	29	2 D	705	89	4 O
233	30	2 E	713	90	4 P
241	31	2 F	721	91	4 Q
249	32	2 G	729	92	4 R
257	33	2 H	737	93	4 S
265	34	2 I	745	94	4 T
273	35	2 K	753	95	4 U
281	36	2 L	761	96	4 V
289	37	2 M	769	97	4 W
297	38	2 N	777	98	4 X
305	39	2 O	785	99	4 Y
313	40	2 P	793	100	4 Z
321	41	2 Q	801	101	5 A
329	42	2 R	809	102	5 B
337	43	2 S	817	103	5 C
345	44	2 T	825	104	5 D
353	45	2 U	833	105	5 E
361	46	2 V	841	106	5 F
369	47	2 W	849	107	5 G
377	48	2 X	857	108	5 H
385	49	2 Y	865	109	5 I
393	50	2 Z	873	110	5 K
401	51	3 A	881	111	5 L
409	52	3 B	889	112	5 M
417	53	3 C	897	113	5 N
425	54	3 D	905	114	5 O
433	55	3 E	913	115	5 P
441	56	3 F	921	116	5 Q
449	57	3 G	929	117	5 R
457	58	3 H	937	118	5 S
465	59	3 I	945	119	5 T
473	60	3 K	953	120	5 U

12mo and

1	1	A	313	27	2 B
5	1*	A 2	317	27*	2 B 2
13	2	B	325	28	2 C
17	2*	B 2	329	28*	2 C 2
25	3	C	337	29	2 D
29	3*	C 2	341	29*	2 D 2
37	4	D	349	30	2 E
41	4*	D 2	353	30*	2 E 2
49	5	E	361	31	2 F
53	5*	E 2	365	31*	2 F 2
61	6	F	373	32	2 G
65	6*	F 2	377	32*	2 G 2
73	7	G	385	33	2 H
77	7*	G 2	389	33*	2 H 2
85	8	H	397	34	2 I
89	8*	H 2	401	34*	2 I 2
97	9	I	409	35	2 K
101	9*	I 2	413	35*	2 K 2
109	10	K	421	36	2 L
113	10*	K 2	425	36*	2 L 2
121	11	L	433	37	2 M
125	11*	L 2	437	37*	2 M 2
133	12	M	445	38	2 N
137	12*	M 2	449	38*	2 N 2
145	13	N	457	39	2 O
149	13*	N 2	461	39*	2 O 2
157	14	O	469	40	2 P
161	14*	O 2	473	40*	2 P 2
169	15	P	481	41	2 Q
173	15*	P 2	485	41*	2 Q 2
181	16	Q	493	42	2 R
185	16*	Q 2	497	42*	2 R 2
193	17	R	505	43	2 S
197	17*	R 2	509	43*	2 S 2
205	18	S	517	44	2 T
209	18*	S 2	521	44*	2 T 2
217	19	T	529	45	2 U
221	19*	T 2	533	45*	2 U 2
229	20	U	541	46	2 V
233	20*	U 2	545	46*	2 V 2
241	21	V	553	47	2 W
245	21*	V 2	557	47*	2 W 2
253	22	W	565	48	2 X
257	22*	W 2	569	48*	2 X 2
265	23	X	577	49	2 Y
269	23*	X 2	581	49*	2 Y 2
277	24	Y	589	50	2 Z
281	24*	Y 2	593	50*	2 Z 2
289	25	Z	601	51	3 A
293	25*	Z 2	605	51*	3 A 2
301	26	2 A	613	52	3 B
305	26*	2 A 2	617	52*	3 B 2

18mo.		
625	53	3 C
629	53*	3 C 2
637	54	3 D
641	54*	3 D 2
649	55	3 E
653	55*	3 E 2
661	56	3 F
665	56*	3 F 2
673	57	3 G
677	57*	3 G 2
685	58	3 H
689	58*	3 H 2
697	59	3 I
701	59*	3 I 2
709	60	3 K
713	60*	3 K 2
721	61	3 L
725	61*	3 L 2
733	62	3 M
737	62*	3 M 2
745	63	3 N
749	63*	3 N 2
757	64	3 O
761	64*	3 O 2
769	65	3 P
773	65*	3 P 2
781	66	3 Q
785	66*	3 Q 2
793	67	3 R
797	67*	3 R 2
805	68	3 S
809	68*	3 S 2
817	69	3 T
821	69*	3 T 2
829	70	3 U
833	70*	3 U 2
841	71	3 V
845	71*	3 V 2
853	72	3 W
857	72*	3 W 2
865	73	3 X
869	73*	3 X 2
877	74	3 Y
881	74*	3 Y 2
889	75	3 Z
893	75*	3 Z 2
901	76	4 A
905	76*	4 A 2
913	77	4 B
917	77*	4 B 2
925	78	4 C
929	78*	4 C 2

16mo.		
1	1	A
17	2	B
33	3	C
49	4	D
65	5	E
81	6	F
97	7	G
113	8	H
129	9	I
145	10	K
161	11	L
177	12	M
193	13	N
209	14	O
225	15	P
241	16	Q
257	17	R
273	18	S
289	19	T
305	20	U
321	21	V
337	22	W
353	23	X
369	24	Y
385	25	Z
401	26	2 A
417	27	2 B
433	28	2 C
449	29	2 D
465	30	2 E
481	31	2 F
497	32	2 G
513	33	2 H
529	34	2 I
545	35	2 K
561	36	2 L
577	37	2 M
593	38	2 N
609	39	2 O
625	40	2 P
641	41	2 Q
657	42	2 R
673	43	2 S
689	44	2 T
705	45	2 U
721	46	2 V
737	47	2 W
753	48	2 X
769	49	2 Y
785	50	2 Z
801	51	3 A
817	52	3 B
833	53	3 C
849	54	3 D
865	55	3 E
881	56	3 F
897	57	3 G
913	58	3 H
929	59	3 I
945	60	3 K

24mo.					
1	1	A	433	19	T
9	1*		441	19*	
17	.....	A 2	449	.....	T 2
25	2	B	457	20	U
33	2*		465	20*	
41	.....	B 2	473	.....	U 2
49	3	C	481	21	V
57	3*		489	21*	
65	.....	C 2	497	.....	V 2
73	4	D	505	22	W
81	4*		513	22*	
89	.....	D 2	521	.....	W 2
97	5	E	529	23	X
105	5*		537	23*	
113	.....	E 2	545	.....	X 2
121	6	F	553	24	Y
129	6*		561	24*	
137	.....	F 2	569	.....	Y 2
145	7	G	577	25	Z
153	7*		585	25*	
161	.....	G 2	593	.....	Z 2
169	8	H	601	26	2 A
177	8*		609	26*	
185	.....	H 2	617	.....	2 A 2
193	9	I	625	27	2 B
201	9*		633	27*	
209	.....	I 2	641	.....	2 B 2
217	10	K	649	28	2 C
225	10*		657	28*	
233	.....	K 2	665	.....	2 C 2
241	11	L	673	29	2 D
249	11*		681	29*	
257	.....	L 2	689	.....	2 D 2
265	12	M	697	30	2 E
273	12*		705	30*	
281	.....	M 2	713	.....	2 E 2
289	13	N	721	31	2 F
297	13*		729	31*	
305	.....	N 2	737	.....	2 F 2
313	14	O	745	32	2 G
321	14*		753	32*	
329	.....	O 2	761	.....	2 G 2
337	15	P	769	33	2 H
345	15*		777	33*	
353	.....	P 2	785	.....	2 H 2
361	16	Q	793	34	2 I
369	16*		801	34*	
377	.....	Q 2	809	.....	2 I 2
385	17	R	817	35	2 K
393	17*		825	35*	
401	.....	R 2	833	.....	2 K 2
409	18	S	841	36	2 L
417	18*		849	36*	
425	.....	S 2	857	.....	2 L 2

THE FOREMAN OR OVERSEER.

GENERAL DUTIES.

OVERSIGHT, vigilant and conscientious, is the price of profit and success. An overseer or foreman of a printing-office should be of more than ordinary capacity, and possess an even and unruffled temper. His conduct should be guided by justice and equity in regard to the interests of the employer and the employed. A strict impartiality should be observed in his treatment of the workmen, and no favouritism should be displayed. He should make himself acquainted with the capacity of the men, and apportion work among them accordingly. Some men are valueless except for plain, straightforward composition; others, distinguished for taste and skill, delight in intricate work or matter requiring ingenuity and delicacy, such as tables, music, and algebra. Put one of the first class on this sort of composition, and he will "botch" it, and earn small wages; while a workman of the latter class will become restive and dissatisfied with plain, solid matter. While dealing justly with the men under his charge, the foreman should see to it that the employer suffers no detriment from negligent or dishonest practices of unconscientious workmen, whether from careless correcting, allowing dropped types to lie upon the floor, or overcharging, or other methods well known in a printing-office. He should be the first and the last in attendance, in order to satisfy himself that every person does his duty in coming and leaving at the proper time.

The office having been thoroughly swept at an early hour, and the type found in any alley having been placed in the stick of the compositor occupying it, the foreman should pass around the room and see that it is immediately distributed, instead of being thrown on the window-frame or table. The type found in the body of the rooms should be sorted out and distributed at once, and not be allowed to accumulate. No pi should be permitted to remain over till the next day. This is an essential point to secure a tidy and well-regulated office.

The foreman should keep himself thoroughly informed of the amount and condition of the materials in the office, not only in gross, but in detail, including every style of type, every variety of accents and peculiar sorts, leads, chasès, furniture, rules, borders, corner-pieces, &c. In this he will be greatly aided by carrying out the good old rule, *A place for every thing, and every thing in its place when not in use*, as well as by keeping a memorandum-book in which every thing should be entered under its proper head for facility of reference.

As a matter of course, he should watch the progress of every job and book, and make sure that they shall be completed within the time contracted for. He should never allow a compositor to have a large take of copy: small takes facilitate expedition, and really tend to the profit of the workmen by bringing an earlier return of letter. He should see to it that every man has his copy closed in proper time, so as not to detain the make-up, and that he passes the make-up without unnecessary delay. As soon as a form or sheet is made up, he should order it to be imposed and a proof pulled, which, with the copy properly arranged, is to be at once handed to the proof-reader. Nor should he allow of any unnecessary delay on the part of the reader, nor on the part of the compositors in correcting the proof when read. When proofs are required by an author, the foreman must forward them promptly to him, and request him to return them at the earliest possible moment. If the proof is not to be sent out, he should have the second reading quickly performed, and the forms prepared for the foundry or the press.

Systematic attention to the above points will tend to the comfort of the overseer, to the advantage of the workmen, and to the profit and satisfaction of the proprietor of the establishment.

The foreman will find a memorandum Press-Book very useful, in which to make entries of the amount of the paper given out by the warehouseman for the various works, the number printed, &c., as well as the name of the pressmen when the work is done on hand-presses.

When given out to wet.	Names of Works.	No.	Signatures.	Date when laid on.	Names of Pressmen.
1865.				1865.	
May 2	Decorative Printing...	1000	11	May 4	Gœtte.
" 4	Quarto Bible	750	18	" 5	Williams.
" 6	American Printer......	1000	20	" 8	Heriot.
" 7	Life of Prescott.........	3000	2	" 9	Smith.

It is commonly the business of the foreman to examine the press revise; in doing which, he will be careful not only to ascertain whether all the corrections marked in the proof are made, but also to look carefully over the sides, head, and bottom of each page. It frequently happens that the folios drop out of the form in lifting it off the imposing-stone; and in leaded matter, letters at the beginning and ends of lines sometimes fall out of place. Before the revise is given to the compositor, the name of the pressman who is to work off the form should be entered in the Press-Book. With foul compositors, he should require a second revise, in order to ascertain if all the corrections have been made which were marked in the first. He should (where there is not a pressman engaged expressly for the purpose, as is the case in houses employing numerous machine-presses) go frequently to the different presses, and examine the work, point out defects, if any, and glance again over the heads, sides, and bottoms of the pages, to see if any thing has been drawn out by the rollers, which may occur from bad justification of the lines, and careless and improper locking up of the form.

An active and conscientious foreman will not be content with merely managing the concerns of the composing-room: he will also see that the business of the warehouse is attended to with regularity and accuracy, and that the warehouseman, errand-boys, and apprentices do their duty.

CASTING OFF COPY.

To cast off manuscript with accuracy and precision, is a task which requires great attention and mature deliberation. The trouble and difficulty are much increased when the copy is not only irregularly written (which is generally the case), but also abounds with interlineations, erasures, and variations in the sizes of paper. At times, so numerous are the alterations and additions as to baffle the skill and judgment of the most experienced calculators of copy. Such an imperfect and slovenly mode of sending works to the press cannot be too strongly censured.

The first step necessary is to take a comprehensive view of the copy, noticing whether it has been written even or has many interlineations, &c., and observing also the number of break-lines, and whether the work be divided into chapters and sub-heads, in order that allowance may be made for them in the calculation. These observations may be noted on a separate piece of paper, to assist the memory and save the trouble of re-examining the manuscript.

This preparation being made, we ascertain the number of words contained in the line by counting several separate lines in various parts of the copy, so that the one we adopt may be a fair average. We then take the number of lines in a page, and multiply by the number of words found in the average line: the quotient we then multiply by the quantity of folios the manuscript copy may contain, and thus we get the amount of words contained in the work, with a tolerable degree of accuracy. The necessary allowances should be made for break-lines, chapters, insertions, &c., according to the observations previously made on the memorandum.

If information has been furnished as to the size of letter the work is to be done in and the width of the page, we make our measure accordingly, and, by composing a few lines of the manuscript copy, we ascertain what number of words will come into each printed line: we then take the length of our page in lines, and multiply the one by the other, thus getting the number of words in the printed page. We divide the whole number of words in the manuscript by the number contained in the printed page: the quotient gives the number of pages the manuscript will make. If too many, the page must be enlarged; if too few, the page must be diminished in width and length. For example:—We take the number of words in a line of manuscript at 20, the lines in a page at 50; we multiply 50 by 20, which will produce 1000 words in a page; we then multiply 1000 by 422, the number of folios in the manuscript, and we find it contains 422,000 words. The work being printed in Pica octavo, 20 ems measure, and each line containing 10 words, each page 40 lines, the case will stand thus:—

MANUSCRIPT.	PRINTED.
50	
20	40
1000	10
422	400)422000 words in MS.
2000	1055 pages.
2000	
4000	Divide
422000 words in MS.	16)1055(65 sheets,
	15 pages.

Another method for casting off copy is the following, as laid down by a predecessor:—

"After having made the measure for the work, we set a line of the letter that is designed for it, and take notice how much copy will come into the line in the stick,—whether less or more than a line of manuscript; and, as it is seldom that neither one nor the other happens, we make a mark in the copy where the line in the stick ends, and number the words that it contains. But, as this is not the safest way for casting off close, we count not only the syllables, but even the letters, that are in a line in the stick, of which we make a memorandum, and proceed to set off a second, third, or fourth line, till a line of copy falls even with a line in the stick; and, as we did to the first line in the stick, so we do to the other, marking on the manuscript the end of each line in the stick, and telling the letters in each, to see how they balance against each other. This being carefully done, we begin counting off, each time, as many lines of copy as we know will make even lines in the stick. For example, if 2 lines of copy make 3 lines in print, then 4 make 6, 6 make 9, 8 make 12, and so on, calling every two lines of copy three in print.

"In like manner we say, if 4 lines make 5, then 8 make 10, and so on, comparing every four lines of copy to five lines in print.

"And in this manner we carry our calculation on as far as we have occasion, either for pages, forms, or sheets.

"The foregoing calculations are intended to serve where a line of print takes in less than a line of copy; and, therefore, where a line of print takes in more than a line of copy, the problem is reversed, and, instead of saying, if 2 lines make 3, we say, in this case, if 3 lines of copy make 2 lines in print, then 6 lines make 4, 9 make 6, 12 make 8, and so on, counting three lines of copy to make two lines in print. In this manner we may carry our calculation to what number of pages, forms, or sheets we will, remembering always to count off as many lines of copy at once as we have found they will make even lines in the stick. Thus, for example, if 5 lines make 7, the progression of 5 is 10, 15, 20, &c., and the progression of 7 will be 14, 21, 28, &c.

"In counting off copy, we take notice of the breaks; and where we judge that one will drive out, we intimate it by a mark of this shape [; and again, where we find that a break will get in, we invert it, thus,]. And to render these marks conspicuous to the compositor, we write them in the margin, that he may take timely notice, and keep his matter accordingly. We also take care to make proper allowance for heads to chapters, sections, paragraphs, &c.

"In examining the state of the copy, we must observe whether it has abbreviations, that we may guard against them in casting off, and allow for them according to the extent of the respective words when written out at length."

The foregoing will convey a sufficient idea as to the best mode of casting off copy; still, these remarks more properly apply to regularly written and thoroughly revised copy. Upon this subject, Smith justly observes,—

"But how often one or more of these requisites are wanting, compositors can best tell; though very few will imagine that among men of learning there should be some who write after such a manner that even those who live by transcribing rather shun than crave to be employed by them: no wonder, therefore, if compositors express not the best wishes to such promoters of printing. But it is not always the capacious genius that ought to be excused for writing in too great a hurry; for sometimes those of no exuberant brains affect uncouth writing, on purpose to strengthen the common notion that *the more learned the man, the worse is his* (hand) *writing;* which shows that writing *well* or *bad* is but a habit with those that *can* write."

HURRIED WORK.

It is sometimes necessary to print pamphlets and other works of a temporary nature in the course of a few hours. When a work of this kind is put in hand, the foreman selects the requisite number of swift and skilful compositors, whose first concern must be to appoint one from among them to make up the matter, and to do every thing which would interfere with the regular business of distributing, composing, and correcting. While they are distributing letter, the *clicker*, or person appointed to manage the work, procures the copy, with all necessary information respecting it, and provides leads, rules, and every other necessary sort. He then draws out the following table:—

Compositors' Names.	Folios of Copy.	Lines Composed.	Memorandums.

In the first column he writes the name of each compositor when he takes copy; and, in the second, the folio of the copy, that he may be able to ascertain instantly in whose hands it lies. In the third column he sets down, opposite to the workman's name, the number of lines composed, as fast as the galleys are brought to him. In the fourth, he inserts such remarks respecting the copy, &c. as may be necessary, and also any circumstances that may occur in the companionship.

When the work is finished, each man's share of lines is readily ascertained, and all disputes are avoided. The publisher may expedite the progress of the work by offering a copy of the book, or some other token, as a premium to the compositor who sets the largest number of ems. The maker-up or clicker usually receives for his compensation the head and foot lines, and two or three cents per thousand, which is deducted from the wages paid to the compositor. Sometimes the compositors work "in pocket," as it is called, or share evenly in the proceeds. This, however, is not a satisfactory mode, and its tendency is to retard the work, as no man will be anxious to do more than his share.

When the compositors are ready for their first taking of copy, it should be given to them in pieces as short as possible, the first two beginning with shorter takes than the others, to prevent delay in the making up. During the time the first take is in hand, the clicker sets the half-head, head-lines, white-lines, and signature-lines, together with notes and other extraneous matter.

When the first person brings his matter, the clicker counts off the number of lines, and inserts them in the table; he then gives him another take of copy, and proceeds with the making up. The same plan is observed with the rest of the compositors. When the first sheet is made up, the clicker lays the pages on the stone, and informs the foreman of it, who will then immediately provide chases and furniture, and the clicker immediately imposes the form.

The proofs should be read at once and given to the clicker to have them corrected. As soon as this is done, he lays up the forms, and gives the proof to the compositor whose matter stands first, who should immediately correct it, then forward it to the next, and so on, till the sheet be corrected; the clicker then locks it up and pulls the second proof, which must be duly forwarded, and the type be locked up finally for press.

The work will now proceed rapidly, provided the compositors stick close to their work and there be no hinderance with respect to letter, &c.: this depends on the good management of the foreman.

If the clicker find that he cannot make up the matter as fast as it is composed, he should call one of the compositors to his assistance, who must be the person last in copy.

COMPANIONSHIPS.

DISPUTES sometimes arise in a printing-office upon trifling as well as important points, which should be settled by a reference to the general custom and usage of the trade. These annoying misunderstandings take place in companionships consisting of several compositors; it is therefore highly desirable that the generally received rules and regulations in this regard should be explicitly laid down for the comfort and government of the compositor.

TAKING COPY.

When the work to be taken in hand is a reprint which is to be followed page for page, a fixed number of pages should be given to each compositor as he comes in turn for copy; or, if the work be in manuscript, an equal average amount should be allowed as a take for each compositor. The foreman should permit none of the hands to have access to the copy, but should deal it out as wanted with perfect impartiality, fat or lean as it may happen to run. Otherwise, a compositor who has an acquaintance with the copy may be tempted to loiter if the next take to be given out be lean, or, if it be fat, to apply for copy before his work in hand is finished. By this course, the foreman will prevent all such sorts of sharp practice, and secure harmony in the companionship. When the foreman gives out copy, he should plainly mark the name of the compositor at the head of the first page of the take if the work be a page-for-page reprint; if it be manuscript, or a reprint in a different measure from that of the copy, he should write the name at the beginning of the first paragraph of the take. Most compositors desire to have a large portion of copy, under the erroneous idea that it will be to their advantage to make up many pages at once. Small takes insure a more rapid execution of the work and bring a quicker return of letter, and so tend to the profit of the hands.

If one of the companionship absents himself, the man next

in order should close his copy, whether it be good or bad, unless the larger portion of it be not set, in which case the person who has the last take must go on with it.

MAKING UP.

The compositor who has the first take on the work proceeds without delay to make it up as soon as he has completed it. Having completed as many pages as his matter will make, he passes the overplus, if less than half a page, with the correct head and folio, to the compositor whose matter follows his, at the same time taking an account of the number of lines loaned; if, on the contrary, the overplus makes more than half a page, he borrows a sufficient number of lines to complete his page; each compositor keeping an account of the number of lines borrowed and loaned. The second compositor, following the same course, passes the make-up to the next in succession; each man passing the make-up in like manner without unnecessary delay.

MAKING UP OF LETTER.

The number of the companionship, if possible, should be determined on at the commencement of the work, to enable all to proceed upon an equal footing. The letter appropriated for the work should be adequate to keep the persons on it fully employed.

If any part of the matter for distribution, whether in chase or in paper, be desirable or otherwise on account of the sorts it may contain, it should be divided equally, or the choice of it thrown for.

When a new companion is put on the work after the respective shares of letter are made up, and if there be not a sufficiency to carry on all the companionship without making up more, he must bring on an additional quantity before he can be allowed to partake of any part of that which comes from the press.

MAKING UP FURNITURE.

The companions in rotation should make up the furniture in turn, the one who has the last matter in the first sheet leading off. Should an odd sheet be wanted, it will be better to throw for the chance of making it up.

IMPOSING AND DISTRIBUTING LETTER.

The person to whose turn it falls to impose must lay up the form for distribution. To prevent disputes, it will be well to prepare a blank form, as follows, which may be filled up as the work proceeds.

Signatures.	Lippincott's Pronouncing Gazetteer.							By whom imposed.
	Maclean.	Armstrong.	Staats.	Devlin.				
B	4	5	4	3				Devlin.
C	3	4	5	4				Maclean.
D								
E								
F								

When the form is laid up, the letter should be divided equally, and, if possible, each person should distribute the matter originally composed by him; by this means, the sorts which may have made his case uneven will return to him. If any man absent himself beyond a reasonable time, his undistributed matter should be divided equally among his companions, and when he returns he may have his share of the next division.

CORRECTING.

The compositor whose matter is first in the proof should lay up the forms on the imposing-stone and correct it; he then hands the proof to the person who follows next. The compositor who corrects the last part of the sheet locks up the forms.

The compositor who has matter in the first and last part, but not the middle of the sheet, only lays up the forms and corrects his matter; the locking up is left to the person who corrects last in the sheet.

A compositor having the first page only of the sheet is required to lay up one form; also to lock up one form if he has but the last page.

If, from carelessness in locking up the form,—viz. the furniture binding, the quoins badly fitted, &c.,—any letters, or even a page, should fall out, the person who locked up the form should repair the damage. But, if the accident occur from bad justification, or from letters *riding* upon the ends of the leads, the loss should fall upon the person to whom the matter belongs.

It is the business of the locker-up to ascertain whether all the pages are of equal length; and, though a defect in this respect is highly reprehensible in the maker-up (whose duty it is to rectify it), yet, if not previously discovered by the locker-up, and an accident happen, he must make good the defect.

The compositor who imposes a sheet must correct the alterations in that sheet. He must also rectify any defect in the register, arising from want of accuracy in the furniture.

Forms sometimes remain a considerable length of time before they are put to press. In this case, particularly in summer, the furniture is likely to shrink, and the pages may fall out. It is therefore the business of the locker-up to attend to it in this respect, or he will be subject to make good any accident which his neglect may occasion.

When forms which have been worked off are ordered to be kept standing, they are considered under the care of the foreman. When they are cleared away, it is to be done in equal proportions by the companionship. During the time any forms may have remained under the care of the foreman, should there have been any alteration as to form or substance which were not made by the original compositors, they are not subject to clear away those parts of the form thus altered.

If the pressmen unlock a form on the press, and any part of it fall out from carelessness in the locking up, they are subject to the loss that may happen in consequence.

The compositor who locks up a sheet takes it to the proof-press, and, after he has pulled a proof, hands it, together with the foul proof, to the reader, and deposits the form in a place appointed for that purpose.

TRANSPOSITION OF PAGES.

Each person in the companionship must lay down his pages properly on the stone for imposition. The compositor whose turn it is to impose looks them over to see if they are rightly placed. Should they, after this examination, lie improperly, and be thus imposed, it will be his business to transpose them; but, should the folios be wrong, and the mistake arise from this cause, it must be rectified by the person to whom the matter belongs. Pages without folios or head-lines, laid down wrongly for imposition, must be rectified by the person who has been slovenly enough to adopt this plan.

RULES AND REGULATIONS TO BE OBSERVED IN A PRINTING-OFFICE.

1. COMPOSITORS are to receive their cases from the foreman or his assistant, free from all pi or improper sorts, with clean quadrate and space boxes, both Roman and Italic, which they are to return to him in equally good condition.

2. When a compositor receives letter, furniture, &c. from the foreman, he is to return any portion not used, in as good state as he received it, the same day.

3. When a case is taken out of the rack, the compositor is to return it into the proper place immediately after he has done with it.

4. No cases to be placed over others, or under the frames, or on the floor.

5. Compositors are to impose their matter and pull a proof as soon as made up, unless directed otherwise, and to correct the proof without unnecessary delay.

6. The proof, when pulled, to be given to the reader, the copy in regular order to accompany the first proof, and the foul proof the second.

7. Compositors are not to leave either type or furniture on the stone.

8. A compositor is not to detain an imposing-stone longer than the nature of the business may require.

9. Head-lines, or other useful materials, on galleys, used during the course of a work, to be cleared away as soon as the work is finished.

10. When a work is done, the compositor, before beginning another work, unless otherwise directed, is to clear away the forms, taking from them the head-lines, white-lines, and odd sorts, as well as the leads and reglets; which, with the furniture of each sheet, and the matter properly tied up for papering, are to be given to the foreman.

11. Sweepings of frames to be cleared away before one

o'clock every day. Matter broken by accident to be cleared away on the same day.

12. The saw, saw-block, bowl, sponge, letter-brush, shears, bellows, &c., to be returned to their respective places as soon as done with.

13. Letter-boards, windows, frames, &c., to be kept free from pi.

14. No person to take sorts from the cases of another without leave, nor hoard useful sorts, not wanting or likely to want them.

15. Compositors employed by the week to work not less than ten hours per day.

16. Unnecessary conversation to be avoided.

CABINET OF CASES.

THE PRESS AND ITS WORKING.

HISTORY OF THE PRESS.

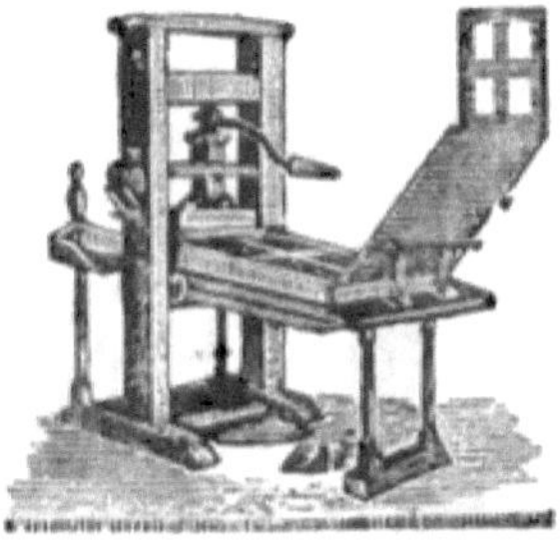

OLD COMMON PRESS.

WHILE poets and orators have expatiated on the glory and power of the press, rulers have exhausted their cunning in attempts to curb and regulate the art of which it is the symbol. Hedged in by arbitrary restrictions, it is not wonderful that printing was long carried on with clumsy implements. The earliest press resembled a screw-press, with a contrivance for running the form of types under the point of pressure. After the impression was taken, the screw was relaxed, and the form withdrawn and the sheet removed.

This rude press continued in general use till 1620, when Willem Jansen Blaeu, at first a joiner and afterward a mathematical instrument maker of Amsterdam, contrived a press in which the bed or carriage was brought under the point of pressure by moving a handle attached to a screw hanging in a beam with a spring, the spring causing the screw to fly back as soon as the impression was given. This movement was afterward effected by means of a double strap or belt, two ends of which were attached to an axle, and the others to opposite ends of the bed. The platen was so small, that two pulls were necessary to print one side of a sheet, and each sheet, therefore, required four pulls to produce a complete impression.

Adam Ramage, who came from Scotland to Philadelphia about 1790, and who for a long time was the chief press-builder in the United States, made some improvements in the old press, one of which was the substitution of an iron bed for the stone one before in use.

About the year 1800, Earl Stanhope contrived a press which obtained much notoriety. It was constructed of iron, and of a size sufficient to print the whole surface of a sheet, and such a combined action of levers was applied to the screw as to make the pull a great deal less laborious to the pressman. But the Stanhope was soon surpassed by the Columbian press, invented by George Clymer, of Philadelphia. Mr. Clymer, as early as 1797, endeavoured to improve the common wooden

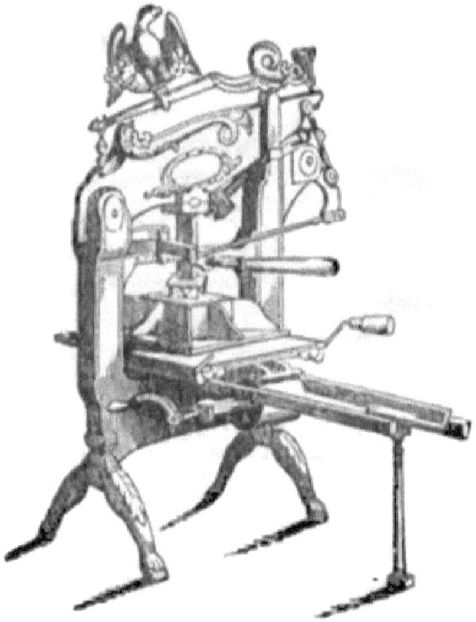

COLUMBIAN PRESS.

press. His next efforts were directed to the production of an iron press, till finally most eminent success was the result of his labours. In beauty, durability, and power, as well as facility of pull, the Columbian press stands, perhaps, unsurpassed. The power in this press is procured by a long bar or handle acting upon a combination of exceedingly powerful levers above the plâten; the return of the handle or levers being effected by means of counterpoises or weights. The powerful command which the leverage enables the workman to exercise is favourable to delicacy and exactness of printing, —his arm feeling, as it were, through the series of levers to the very face of the types. The inventor removed to England in 1817, and introduced the press there, where it was held in high estimation. In the United States, presses of simpler construction have displaced the imposing Columbian press,— the first of which was invented by PETER SMITH, of New York, and the latest is SAMUEL RUST'S Washington press,

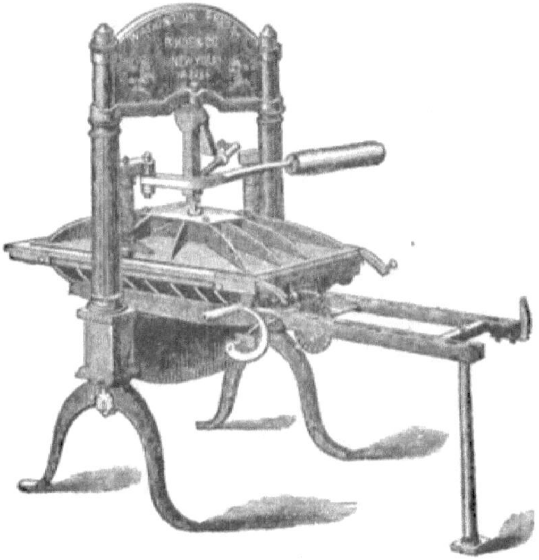

WASHINGTON PRESS.

which has secured general approbation and adoption, as being more simple and cheaper, if not more effective, than the Columbian press. Hand-presses are now restricted to country papers of small circulation, and to book-offices devoted to extra fine printing.

Most books, however, are printed on the bed-and-platen power-press invented by ISAAC ADAMS, of Boston,—the only machine-press yet discovered that is capable of producing fine work and exact register. It will give from six to eight thousand impressions per day. The Adams press is made with two, four, or six inking rollers, as may be ordered. As the platen rolls off and leaves the bed entirely exposed, forms

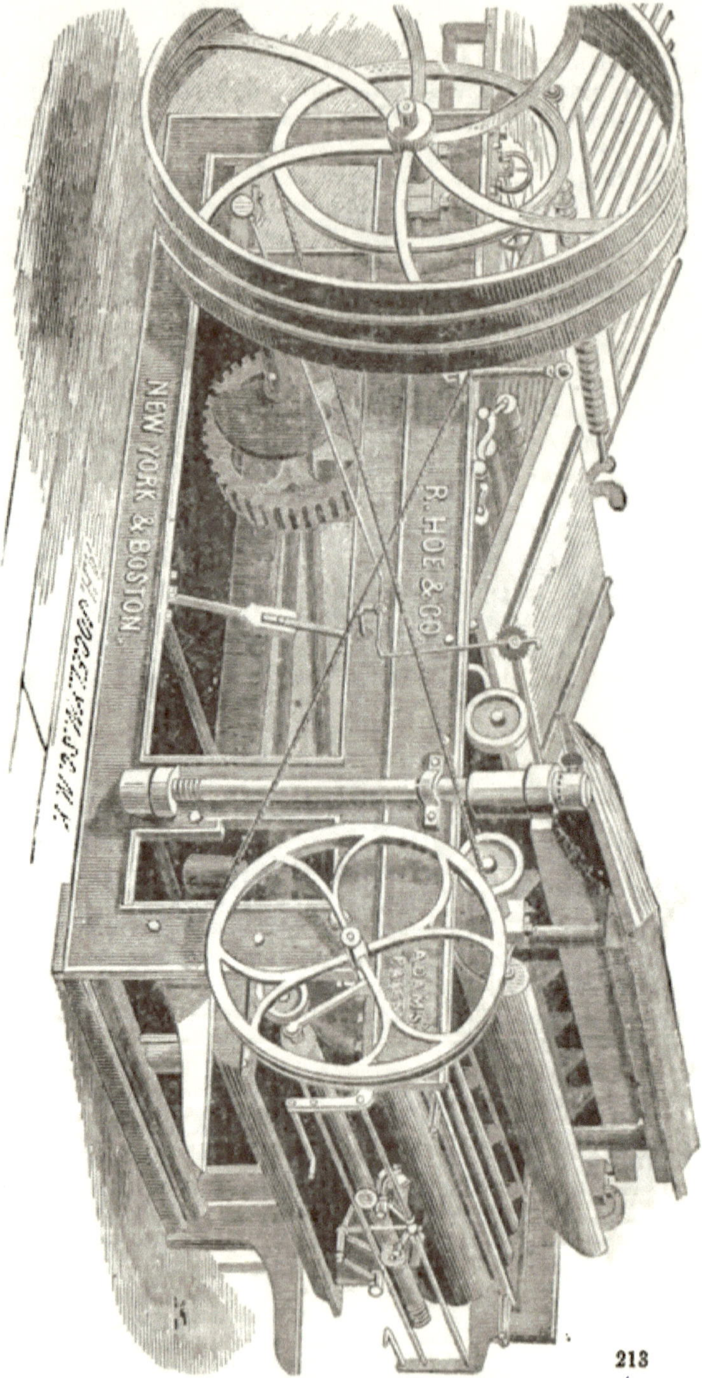

ADAMS PRESS.

can be made ready with great facility. The sheets are taken from the feed-board by fingers, and, after being printed, are laid in a pile by a self-acting sheet-flyer. R. Hoe & Co., who own the patent right, are the sole manufacturers of this press.

The Cylinder press, which may be run at a much higher rate of speed than the bed-and-platen machine, was of earlier invention. FREDERICK KÖNIG, a Saxon, early in the present century turned his attention to cylinder printing, and was so successful that on November 28, 1814, the London *Times* announced the fact that the number issued on that day had been printed by machinery propelled by steam. The first suggestion of a cylinder press is due, nevertheless, to William Nicholson, an Englishman, who in 1790 took out a patent for such a machine, but which was never acted on. Isaiah Thomas says that a Dr. Kinsley, of Connecticut, afterward produced a press varying somewhat from Nicholson's.

In 1818, Applegath and Cowper made important improvements in König's press, which greatly enlarged its field of usefulness. This machine, with various modifications and improvements, is in general use in Europe and America, for newspapers of moderate circulation, and job-work in which accurate register is not indispensable; though by some of the machines made by R. Hoe & Co.* and A. B. Taylor & Co. of New York, exact register is very nearly attainable.

---

* We are indebted to a friend for the following sketch of the origin, progress, and present condition of the world-famous house of R. Hoe & Co.

Robert Hoe, the founder of the present house of R. Hoe & Co. of New York, was born at Hose, in Leicestershire, England, in 1784. His father was a well-to-do farmer in that pleasant, sequestered district; but, as the family was large, Robert was apprenticed to a carpenter in a neighbouring town. His attention was early attracted and his mind impressed by the prosperity of the people of the United States; and, being a republican at heart, and conscious that the institutions of his own country presented almost insurmountable obstacles to the advancement of the working classes, he purchased his indentures from his employer, and in 1803 emigrated to New York.

On his arrival he made the acquaintance of Grant Thorburn, who, becoming interested in him, received him into his family, and with great kindness nursed him with his own hands through an attack of the yellow fever, which was then raging in the city. He soon established himself in his trade, and, by his industry, integrity, and enterprise, became advantageously known to the public. At the age of twenty he married the daughter of Matthew Smith, of Westchester Co., New York, by whom he had three sons and six daughters. For a time he was in partnership with his brother-in-law, Matthew Smith, Jr., a carpenter and

The invention of steam printing presses rendered books and periodicals so cheap that the progress of knowledge was amazingly accelerated; and soon the achievements of the cylinder press proved unequal to the work of printing the

---

printers' joiner, who, on their separation, associated with himself his brother, Peter Smith, who was educated at Yale College, and was the inventor of the well-known hand-press bearing his name.

On the decease of these two brothers, Robert Hoe, in 1823, succeeded to the business, which was then in its infancy, giving employment to only a handful of men, and being conducted in the middle of the block bounded by Maiden Lane, Pine, William, and Pearl Streets, in some old buildings to which access was gained by an alley running from Maiden Lane to Pine Street. Here the business, under the style of Robert Hoe & Co., grew rapidly; but the extension of Cedar Street made necessary its removal to the present location in Gold Street. About this time, the flat-bed cylinder press, for newspaper printing, was introduced into England; and Mr. Hoe sent an intelligent mechanic there to examine it, and it was soon brought into use here, with valuable improvements. In 1832, Mr. Hoe's failing health obliged him to relinquish the business to his eldest son, Richard M. Hoe, and Matthew Smith, son of his first partner. In the following year he died. Shortly after his decease, the firm erected extensive buildings in Broome Street, in the eastern part of the city, where the greater part of their manufacturing has since been carried on. They also commenced making cast-steel saws, which had previously been exclusively imported from England; and this branch has steadily increased in importance.

Matthew Smith, a man of uncommon ability and business talents, died in 1842. The business was then continued by Richard M. Hoe, with his two brothers, Robert Hoe and Peter Smith Hoe,—the eldest, as before, taking charge of the mechanical department, in which his industry and fertility of invention are attested by the number and value of his patents. In 1837, he patented here and in England his method of grinding circular saws, by which the thickness of any part of a saw can be regulated with accuracy. In 1846, he brought out the so-called "Lightning Press," or Type-Revolving Printing Machine, described in the text,—the greatest innovation on the routine of the printing craft since the days of Gutenberg. This press has entirely superseded all others for fast printing, being at present in use in all the principal offices, not only in this country, but in England, Scotland, Ireland, and Australia, with a constantly increasing demand.

In 1858, the firm purchased of Isaac Adams, of Boston, Massachusetts, his entire patent-rights, together with his establishment for the manufacture of his bed-and-platen book printing presses, and various machines for binders' use, which they continue to conduct there, though with increased facilities and many improvements. Their works in different places now cover thirty-five city lots, or about two acres, and give employment to nearly six hundred hands. The office and warerooms of the house in England are at 13 Salisbury Square, Fleet Street, London, one of the partners, being a resident in that city, attending to the business there.

DOUBLE CYLINDER PRINTING MACHINE.

enormous editions of some of the leading newspapers of the world; and an invention to meet the exigency was successfully made by Col. RICHARD M. HOE, of New York, in the

HOE'S TYPE-REVOLVING MACHINE—FRONT VIEW.

SIDE VIEW.

Type-Revolving Printing Machine, of which we give two engravings. It is, as its name indicates, on the rotary principle; that is, the form of type is placed on the surface of a horizontal revolving cylinder of about four and a half feet in diameter. The form occupies a segment of only about one-fourth of the surface of the cylinder, and the remainder is used as an ink-distributing surface. Around this main cylinder and parallel with it, are placed smaller impression cylinders varying in number from four to ten, according to the size of the machine. The large cylinder being put in motion, the form of types is carried successively to all the impression cylinders, at each of which a sheet is introduced and receives the impression of the types as the form passes. Thus, as many sheets are printed at each revolution of the main cylinder as there are impression cylinders around it. One person is required at each impression cylinder to supply the sheets of paper, which are taken at the proper moment by fingers or grippers, and after being printed are carried out by tapes and laid in heaps by means of self-acting flyers, thereby dispensing with the hands required in ordinary machines to receive and pile the sheets. The grippers hold the sheet securely, so that the thinnest newspaper may be printed without waste.

The ink is contained in a fountain placed beneath the main cylinder, and is conveyed by means of distributing rollers to the distributing surface on the main cylinder. This surface being lower, or less in diameter, than the form of types, passes by the impression cylinder without touching. For each impression there are two inking rollers, which receive their supply of ink from the distributing surface of the main cylinder: they rise and ink the form as it passes under them, after which they again fall to the distributing surface.

Each page of the paper is locked up on a detached segment of the large cylinder, which constitutes its bed and chase. The column-rules run parallel with the shaft of the cylinder, and are, consequently, straight; while the head, advertising, and dash rules are in the form of segments of a circle. The column-rules are in the form of a wedge, with the thin part directed toward the axis of the cylinder, so as to bind the types securely. These wedge-shaped column-rules are held down to the bed by tongues projecting at intervals along their length, which slide in rebated grooves cut crosswise in the face of the bed. The spaces in the grooves between the column-rules are accurately fitted with sliding blocks of metal even with the surface of the bed, the ends of which blocks are cut away underneath to receive a projection on the sides of the tongues of the column-rules. The form of type is locked up in the bed by means of screws at the foot and sides, by which the type is held as securely as in the ordinary manner upon a flat bed,—if not even more so. The speed of these machines is limited only by the ability of the feeders to supply the sheet.

This machine was first used by the *Public Ledger* of Philadelphia, and was afterward adopted by the leading newspapers of that city and New York, as well as of the chief cities of France and England.

BULLOCK'S PATENT ROTARY PERFECTING PRESS.

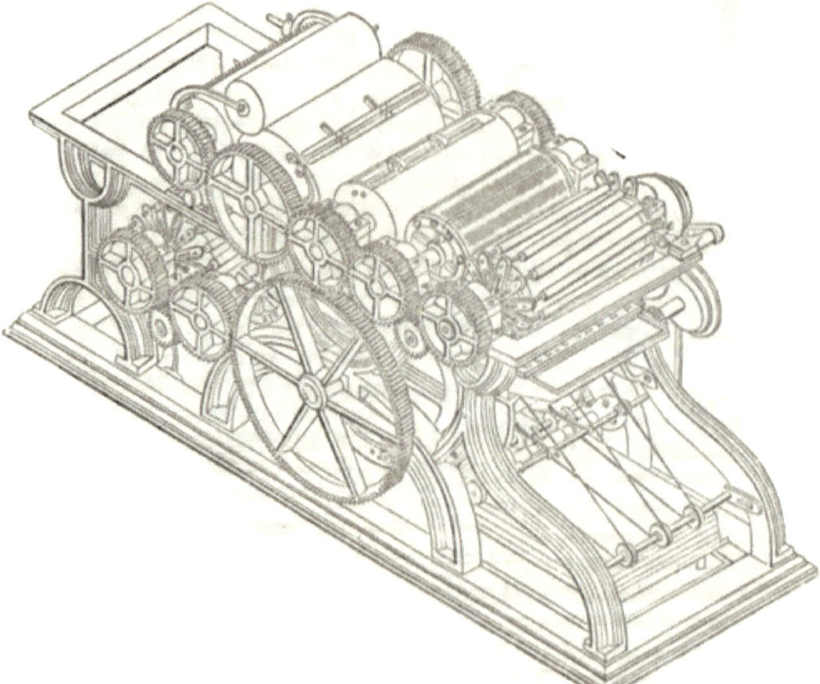

THIS American machine has been devised by William Bullock for printing on a continuous sheet of dampened paper, unrolled from a cylinder, which, before it leaves the machine, receives impressions on both sides. Before the impression is made (from stereotype plates bent around small horizontal cylinders), the sheets are ingeniously cut of the proper dimensions, and when printed are deposited in a single pile. In the office of the *Philadelphia Inquirer* there are two single presses running nine thousand perfect copies per hour each, and one double press producing fifteen thousand copies per hour.

PATENT RAILROAD-TICKET PRINTING AND NUMBERING PRESS.

THE invention of this ingenious machine is due to the sagacity and enterprise of those tasteful and excellent printers, SANFORD, HARROUN & Co., of New York. With a liberal expenditure of money, they employed a pressman, a compositor, and an engineer, to work out their ideas, until complete success rewarded the combined effort.

The machine not only prints, but at the same operation numbers consecutively, tickets and coupons of every size and pattern, which are also indented or cut apart by the machine as fast as printed. Ten local tickets, of ten different forms if desired, or coupon tickets twenty-six inches in length, can be printed at one operation of the press. The average rate of speed is about fourteen hundred impressions an hour,—equivalent to fourteen thousand tickets.

JOB PRESSES.

THE invention of machines for printing small work elegantly as well as swiftly is of vast advantage to the printer, and has greatly increased the jobbing department of typography. Here, as in other matters, American ingenuity has taken the lead of all nations; and the presses invented by Ruggles, Adams, Hoe, Wells, Gordon, and Degener—not to mention numerous other inventors—defy competition. The Ruggles presses formerly commanded the trade; but the beautiful machines of George P. Gordon, a man of decided genius, have displaced them in public estimation. Seth Adams's machine, made by R. Hoe & Co., is highly esteemed.

Gordon's Firefly press is unique, and requires a so-called endless card board, which it prints and cuts of the required shape as it goes, at the rate of about ten thousand per hour. This is not in general use; but his Eighth Medium, Quarto Medium, and Half Medium Franklin presses, here shown, have achieved a high reputation for expedition and excellent performance.

GORDON'S FRANKLIN PRESS.

HOE'S BED-AND-PLATEN JOBBER.

The opposite cut represents a new jobber lately introduced by R. Hoe & Co. of New York, which, judging from the well-known qualities of all their manufactures, will doubtless prove to be a good machine. It is said to be simple in construction, and exceedingly strong and durable.

SETTING UP A WASHINGTON PRESS.

ALL the connecting parts being marked, or indented by points, if these be observed carefully, the press may be put together without difficulty.

After setting the frame upon its legs, and putting on the ribs and bed, lay the platen on the bed, placing under it two bearers about type high. Then put the springs in their places, and the nuts over them, and pass the suspending-rods through them, observing to place the rods so that the number of indentations on them correspond with those on the platen. Give the nuts two or three turns, then run in the bed so as to bring the platen under the rods, and screw them fast to the platen; after which, put in the bar-handle, standard, and lever (or wedge and knees, if a Smith press). Turn the nuts on the suspending-rods, so as to compress the springs just enough to give the platen a quick retrograde motion, observing at the same time to get the surface of the platen parallel with the surface of the bed.

After having put the press together and levelled it by means of a spirit-level, be particular not to raise the end of the ribs by the gallows, but let it go under rather loose, which will have a tendency to make the bed slide with more ease on the ribs.

SETTING UP THE ROLLER-STAND.

THE roller-stand containing the distributing cylinder should be regulated to the height of the press, bringing the shelf or bridge even with the corner irons, and of sufficient distance from the bed to allow it to run clear; the stand should then be firmly braced, as the constant turning of the rounce is very apt to loosen it; meanwhile being cautious to observe that the rounce, in its revolutions, does not come in contact with the frame of the tympan when up. The position of the distributing cylinder should be sufficiently high to allow the two composition rollers, at least one inch apart, to rest on its top without danger of touching the shelf or bridge in front. It is advantageous to nail two narrow strips of sole leather on the face of the shelf, about eight or ten inches from each end, which, acting as bearers, cause the rollers to pass very smoothly over them.

The roller-handle while in use should lie in a horizontal position, the back end being supported by a bar of wood or iron running parallel with the distributing cylinder. There should be a notch, or hook, about two inches from the end of the handle, to catch on the wooden supporter, to prevent the rollers from jumping forward while distributing or changing. It is also necessary to have a back-board for the end of the roller to strike against in coming off the form, to prevent the rollers from falling backward.

The ink-block is placed about five or six inches to the right of the roller-handle, and about on a level with it. It is furnished with the ink-slice, and a brayer, or a small roller about four or five inches long, and of the same circumference with the larger rollers, being cast in the same mould.

COMPOSITION ROLLERS.

Put the glue in a bucket or pan, and cover it with water; let it stand until more than half penetrated with water, taking care that it shall not soak too long, and then pour it off and let it remain until it becomes soft, when it will be ready for the melting kettle. This is a double vessel, like a glue kettle. Put the soaked glue into the inner vessel, and as much water in the outer boiler as it will contain when the inner vessel is placed in it. When the glue is all melted (if too thick, add a little water), the molasses may be slowly poured into it, and well mixed with the glue by frequent stirring. When properly prepared, the composition does not require boiling more than an hour. Too much boiling candies the molasses, and the roller, consequently, will be found to lose its suction much sooner. In proportioning the material, much depends upon the weather and temperature of the place in which the rollers are to be used. Eight pounds of glue to one gallon of sugar-house molasses, or syrup, is a very good proportion for

MELTING KETTLE.

summer, and four pounds of glue to one gallon of molasses for winter use.

For *hand-press rollers* more molasses should be used, as they are not subject to so much hard usage as *cylinder-press rollers*, and do not require to be as strong; for the more molasses that can be used the better will be the roller. Before pouring a roller, the mould should be perfectly clean, and well oiled with a swab, but not to excess, as too much oil makes the face of the roller seamy and ragged. The end pieces should then be oiled, and, together with the cylinder, placed in the mould, the upper-end piece being very open, to allow the composition to pass down between the interior of the mould and the cylinder. The cylinder must be well secured from rising, before the composition is poured in, by placing a stick upon the end of it, sufficiently long to reach above the end of the mould, and be tied down with twine. The composition should be poured very slowly, and in such a manner as to cause it only to run down *one side* of the cylinder, allowing the air to escape freely up the other.

If the mould is filled at night, the roller may be drawn the next morning; but it should not be used for at least twenty-four hours after, except in very cold weather.

To determine when a roller is in order for working, press the hand gently to it: if the fingers can be drawn lightly and smoothly over its surface, it may be said to be in order; but should it be so adhesive that the fingers will not glide smoothly over its surface, it is not sufficiently dry, and should be exposed to the air.

Rollers should not be washed immediately after use, but should be put away with the ink on them, as it protects the surface from the action of the air. When washed and exposed to the atmosphere for any length of time, they become dry and skinny. They should be washed about half an hour before using them. In cleaning a *new* roller, a little oil rubbed over it will loosen the ink: and it should be scraped clean with the back of a case-knife. It should be cleaned in this way for about one week, when *ley* may be used. New rollers are often spoiled by washing them too soon with ley. Camphene may be substituted for oil; but, owing to its combustible nature, it is objectionable, as accidents may arise from its use.

Mr. Hansard, an eminent English printer, says, "Take glue, made from the cuttings of parchment or vellum, fine green molasses, pure as from the sugar-refiners, and a small quantity of the substance called Paris-white, and you will have every ingredient requisite for good composition. The proportion as follows:—

Glue, 2 lbs. Molasses, 6 lbs. Paris-white, ½ lb.

Put the glue in a little water for a few hours to soak; pour off the liquid; put the glue over the fire, and when it is dissolved add the molasses, and let them be well incorporated together for at least an hour; then, with a very fine sieve, mix the Paris-white, frequently stirring the composition. In another hour, or less, it will be fit to pour into the mould."

COVERING TYMPANS.

THE tympans are generally covered with parchment. They should be of an even thickness, and about two inches and a half wider and three inches longer than the tympans. Tympans have been sometimes covered with linen, which, on account of its evenness, would answer the purpose; but it is so apt to stretch, that the tympans become slack in a short time, and bag (as it is termed), and thus slur the impression. Silk is excellent for fine work.

The pressman spreads as much good paste on the edges of the skin as will cover the frame of the tympan, which is also well pasted. He then lays the skin on the inner side of the frame, with the flesh side to face the type, and draws it regularly, as tight as possible, on all sides. The part of the skin that comes on the grooves of the tympan which receive the point-screws, is cut and wrapped round the inside edge of the grooves, which admits a free passage for the screws. After having fastened the skin on the sides of the tympan, he draws it toward the joints which receive the frisket, and with a knife cuts across these joints to let them through the skin; he then puts the frisket-pins through the parchment, and makes that end of the tympan fast. He next proceeds to the lower joints, and brings the skin as tight as he can round that part of the tympan. The point-screws and duck-bill are then put on, which prevent the skin from starting. The inner tympan, or drawer, is covered in the same manner. To prevent their warping when the skin begins to draw, pieces of furniture or wood of any kind should be placed across the centre till they are perfectly dry.

The skins are put on either wet or dry: if dry, they should be afterward well wet, which will make them give somewhat; but when they dry they will contract, and by this means will be rendered much tighter than they would be if put on wet.

WETTING PAPER.

THE size of the wetting-trough should be about two inches longer and wider than the largest-sized paper, folded, that is to be wet in it, and about six inches deep. It should have a cover with hinges on the left side, that the cover may fall over on that side, and, resting horizontally, serve for a shelf to lay the paper upon previous to wetting it.

Having received a certain amount of paper from the warehouseman, the pressman lays one heap on the shelf attached to the wetting-trough, laying the first token across the heap with the back of the quires toward his right hand, that he may know when to turn the token-sheet, and that he may more readily catch at the back of each quire with that hand, for the purpose of dipping it. He then places the paper-board with its breadth before him on his right, on a table, laying a wrapper or a waste sheet of paper on the board, to prevent soiling the first sheet of the heap.

He then takes a quire by the centre of the back with his right hand, and the edge of it in his left, and, closing his hands a little, that the quire may bend downward between his hands, he dips the back of the quire into the left-hand side of the trough, and, relinquishing his hold with the left hand, draws the quire briskly through the water with his right. As the quire comes out, he quickly catches the edge of it again in his left hand, and brings it to the heap, and, by lifting his left hand, bears the under side of the quire off the paper previously laid down, till he has placed the quire in an even position; if the paper be weak and spongy, he draws the quire through the water quickly; if strong and stubborn, slowly. To place the quire in an even position, he lays the back of it exactly upon the open crease of the former, and then lets the side of the quire in his left hand fall flat down upon the heap, and, discharging his right hand, brings it to the edge of the quire, and, with the assistance of his left thumb, still in its first position, opens or divides either a third or a half of the quire, according to the quality of the paper; then, spreading the fingers of his right hand as much as he can through the length of the quire, turns over his opened division of it upon his right-hand side of the heap.

A different process must be used in the wetting of drawing and plate papers. These papers are usually sent in quite flat; that is, not folded into quires or half-quires. The best method of wetting these papers is to use a brush, such as is called a banister brush; and, instead of dipping the paper into the trough, he lays it on the paper-board by the side of the trough, and, dipping the brush into the water, he shakes it gently over the whole surface, to give an equal degree of moisture to all parts; and then proceeds as before described. The drawing-paper, being very hard-sized in the making, will require the brush, and much water, three, four, or even five times a quire; while the plate-paper should have as little water as it is possible to give it, so as to cover it all over; and twice a quire will often be too much. This same mode must also be adopted in wetting paper of extraordinary dimensions.

Having wet his first token, he doubles down a corner of the upper sheet of it on his right hand, so that the farther corner may be a little toward the left of the crease in the middle of the heap, and the other corner may hang out on the near side of the heap about an inch and a half. This sheet is called the *token-sheet*, being a mark for the pressman, when he is at work, to show how many tokens of that heap are worked off.

Having wet the whole heap, he lays a wrapper, or waste sheet of paper, upon it; then, three or four times, takes up as much water as he can in the hollow of his hand, and throws it over the waste sheet, to moisten and soak downward into the wet part of the last division of the quire; after which, he places in the heap the label which the warehouseman must always furnish for each heap, and upon which are written the title of the work and the date of wetting, one-half hanging out so as to be easily read.

The paper should be pressed for twelve hours, and then carefully turned by each three or four sheets, so that no lift be relaid in the same position with respect to the adjoining lift; at the same time, every fold and wrinkle must be carefully rubbed out by the action of the hand, so that nothing but a flat and even surface shall remain; the heap should then be pressed for about twenty-four hours in a screw-press, and it will be in good order for working.

The wetting of paper must, in all cases, depend entirely upon its fabric; and, since the printer has seldom the choice of the paper, it will require all his skill and patience to adapt his labours to the materials upon which he is to work. The texture of the paper must be suited to the fineness and tenacity of the ink. To attempt doing fine work upon common paper is lost labour. A paper to take the best ink must be made entirely of linen rags, *and not bleached by chemicals.* A fine hand-made paper, fabricated a sufficient time to get properly hardened, and well and equally saturated with size, so as not to imbibe more water in one part of the dip than in another, nor resisting the water like a duck's back, is most suitable for fine printing.

BLANKETS.

WOOLLEN blankets are unnecessary when a book is printed from new type. Nothing more should be used than a sheet or two of paper, as in fine work only the face of the type should show in the impression. But when the types or plates are worn and rounded, fine cassimere or broadcloth should be used in the tympan. In this, as in all matters connected with artistic typography, the pressman must display good judgment and discretion.

MAKING READY A FORM.

BEFORE a form is laid on the press, the pressman should carefully wipe the bottom of the type and the bed perfectly clean; for, if a particle of sand remain on it, it will cause a type or two to rise, and not only make a stronger impression, but probably injure the letters.

An octavo form should be laid on the press with the signature-page to the left hand, or nearest the platen; a duodecimo, or its combinations, with the signature at the right hand, or nearest the tympan. The form should be laid under the centre of the platen, and properly quoined up. The tympan is then laid down, and wet if necessary, and paper or blanket put in. It was formerly customary to wet the tympans for all works, and even jobs of almost every description; but, since the introduction of fine printing, and particularly iron presses, the custom is wellnigh banished, excepting for very heavy forms, composed with old letter, which, of course, require more softness to bring them off. After the inner tympan or drawer is put in, it is fastened with the hooks for that purpose, which serve to keep it from springing out. The tympan being lifted up, a sheet of the paper to be worked is folded in quarto, and the short crease is placed over the middle of the grooves of the short cross, if it lie in the centre of the form, as in octavo. In a form of twelves, the paper is folded in thirds, and the long crease placed in the middle of the long cross, and the short cross over the grooves. The sheet lying evenly on the form, the tympan is brought down, and a gentle pull will cause the paper to adhere, when it should be pasted to the tympan and fully stretched. The points are next screwed to the tympan, for large paper short-shanked points being used, and long-shanked for small paper. In twelves, the points must be placed at precisely equal distances from the edge of the paper. In octavo, the off-point may be a little larger than the near one, as it enables the pressman to detect a turned heap when working the reiteration or second side.

When a press is continued upon the same work, the quoins on the off-side of the bed may remain and serve as gauges for the succeeding forms; for, if the chases are equal in size, the register will be almost, if not quite, perfect.

The following operations are comprised in the term of making ready the form:—

1. The frisket should be covered with stout even paper, in the manner described for putting on parchment, the paper being carefully placed on the inside of the frame so as to lie close to the tympan, and to confine the sheet in its place when laid on for printing. When the paste is dry, the frisket is put on the tympan, and, after inking the form, an impression pulled upon it. The frisket is then taken off and laid on a board, or on the bank, and the impression of the pages cut out with a sharp knife about a Pica em larger than the page. After being replaced on the tympan, it is advisable to put a few cords across, to strengthen the bars of paper, and to keep the sheets close to the tympan. When the margin is too small to admit bars of paper, it is necessary to work with cords only.

2. The form should be examined, to see that it is properly

locked up and planed down; that no letters or spaces lie in the white lines of the form, nor between the lines in leaded matter.

3. White pages which occur in a form must not be cut out; but, if the page be already cut out, a piece of paper must be pasted on the frisket, to cover the white page in the form, and a bearer put on to keep the adjoining pages from having too hard an impression. Some pressmen use reglets, others furniture cut to a proper height, and a third class adopt cork, which, from its elasticity, is very useful. Spring bearers, made of hard paper rolled up, are also employed to guard the sides and bottoms of light and open pages, when there is an inclination to slur.

4. The pressman must examine whether the frisket bites; that is, whether it keeps off the impression from any part of the pages.

5. He must consider whether the catch of the frisket stands either too far forward or backward: if forward, he may be much delayed by its falling down, and, if backward, it will come down too slowly, and thus retard the progress of the work, and not unfrequently cause the sheet to slip out of its proper place. He must, therefore, place the catch so that the frisket may stand a little more than perpendicularly backward, that, when lightly tossed up, it may just stand, and not come back.

6. He must fit the gallows so that the tympan may stand as much toward an upright as he can; because it is the sooner let down upon the form and lifted up again. But yet he must not place it so upright as to prevent the white sheets of the paper from lying secure on the tympan.

7. The range of the paper-bank should not stand at right angles with the bed of the press; but the farther end of the bank should be placed so that the near side may make an angle of about seventy-five degrees with the near side of the bed.

8. The heap of paper should be set on the horse on the near end of the paper-bank, near the tympan, yet not touching it. The uppermost or outside sheet should be laid on the bank; and the pressman then takes four or five quires off his heap, and shakes them at each-end, to loosen the sheets, till he finds he has sufficiently loosened or hollowed the heap. Then, with the nail of his right-hand thumb, he draws or slides forward

the upper sheet, and two or three more commonly follow gradually with it, over the hither edge of the heap, to prepare those sheets ready for laying on the tympan.

9. He must next pull a *revise* sheet, which must be sent up to the overseer for a final revision, and for examining whether any letters have dropped out of the form in putting it on the press, &c.

10. While the sheet is undergoing a revision, the pressman should proceed to *make register*, if half-sheet-wise, which is done by pulling a waste sheet, and turning it (without inking, as the sheets may afterward be used for *slip* sheets), being particular not to stretch the point-holes in the least, or to draw the hand along the sheet in leaving it. In making register, the points must be knocked up or down in such a direction as will bring the first impression under the last, knocking the point only half the distance apparent on the sheet. If register cannot be made with the points, the difficulty must then be either in the furniture, the length of the pages, or in the springing of the cross-bars, from the forms being locked up by careless compositors, who commence at one quarter of the form, and lock it up tightly, and so go around, instead of gently tapping it at opposite sides till the whole is equally tightened. In locking up a form, the quoins at the feet should be gently struck first, to force up the pages and prevent their hanging; but, in unlocking, the side quoins must be first slackened.

Altering the quoins will not make good register, when the compositor has not made the white exactly equal between all the sides of the crosses. The pressman, therefore, will ascertain which side has too much or too little white, and, unlocking the form, will take out or put in as many leads or reglets as will make good register.

PULLING.

In taking a sheet off the heap, the pressman places himself almost straight before the near side of the tympan, but nimbly twists the upper part of his body a little backward toward the heap, the better to see that he takes but one sheet off. This he loosens from the rest of the heap by drawing the back of the nail of his right thumb quickly over the bottom part of the heap, and, receiving the near end of the sheet with his left-hand fingers and thumb, catches it by the farther edge with his right hand, about four inches from the upper corner of the sheet, and brings it swiftly to the tympan: having the sheet thus in both his hands, he lays the farther side and two extreme corners of the sheet down even upon the farther side and extreme farther corners of the tympan-sheet. In the reiteration, care should be taken to draw the thumb on the margin, or between the gutters, to avoid smearing the sheet. The sheet being properly laid on, he supports it in the centre by the fingers of the left hand, while his right hand, being disengaged, is removed to the back of the ear of the frisket, to bring it down upon the tympan, laying at the same moment the tympan on the form. He then, with his left hand, grasps the rounce, and quickly runs the form under the platen; and, after pulling, he gives a quick and strong pressure upon the rounce, to run the carriage out again. Letting go the rounce, he places the fingers of his left hand toward the bottom of the tympan, to assist the right hand in lifting it up, and also to be ready to catch the bottom of the sheet when the frisket rises, which he conveys quickly and gently to the catch: while it is going up, he slips the thumb of his left hand under the near lower corner of the sheet, which, with the aid of his two forefingers, he raises, the right hand at the same time grasping it at the top in the same manner. Lifting the sheet carefully and expeditiously off the points, and nimbly twisting about his body toward the paper-bank, he carries the sheet over the heap of white paper to the bank, and lays it down upon a waste sheet or wrapper; but, while it is coming over the white-paper heap, though he has the sheet between both his forefingers and thumbs, yet he holds it so loosely that it may move between them as on two centres, as his body twists about from the side of the tympan toward the side of the paper-bank.

When the pressman comes to a token-sheet, he undoubles it, and smooths out the crease with the back of the nails of his right hand, that the face of the letter may print upon smooth paper: and, being printed off, he folds it again, as before, for a token-sheet, when he works the reiteration.

Having worked off the white paper of a form of twelves, he places his right hand under the heap, and, his left hand supporting the end near him, turns it over on the horse, with the printed side downward. If the form be octavo, he places his left hand under the heap, supporting the outside near end with his right hand, and turns it one end over the other. All turning of the paper for reiteration is treated in one of these modes. In performing this operation, he takes from the heap only as much at once as he can well handle without disordering the evenness of the sides of the paper.

Having turned the heap, he proceeds to work it off, as before described, except that with the left hand he guides the point-holes over the points, moving the sheet with the right hand, more or less, to assist him in so doing. The token-sheets, as he meets with them, he does not fold down again.

RULES AND REMEDIES FOR PRESSMEN.

About every five or six sheets a small quantity of ink should be taken; yet this rule is subject to some variation from the nature of the work and quality of the ink. A form of large type or solid matter will require ink to be taken more frequently, and a light form of small type less frequently. During the intervals in which the roller-boy is not employed in brayering out or taking ink, he should be almost constantly engaged in distributing or changing his rollers. He should invariably take ink on the back roller, as it will the sooner be conveyed to the other roller, and, consequently, save time in distributing. When, through carelessness, too much ink has been taken, it should be removed by laying a piece of clean waste paper on one of the rollers, and distributing them till the ink is reduced to the proper quantity.

If letters, quadrates, or furniture rise up and black the paper, they should be put down, and the quarter locked up tighter.

If any letters are battered, the quarter they are in must be unlocked, and perfect ones put in by the compositor.

When bearers become too thin by long working, they should be replaced by thicker ones.

When the form gets out of register,—which will often happen by the starting of the quoins which secure the chase,—it must be immediately put in again, as there can scarcely be a greater defect in a book than a want of uniformity in this particular.

If picks, produced by bits of paper, composition, or film of ink and grease or filth, get into the form, they must be removed with the point of a pin or needle; but if the form is much clogged with them, it should be well rubbed over with clean ley, or taken off and washed: in either case, before the pressman goes on again, it should be made perfectly dry by pulling several waste sheets upon it, in order to suck up the water deposited in the cavities of the letter.

The pressman should accustom himself to look over every sheet as he takes it off the tympan: he will thus be enabled not only to observe any want of uniformity in the colour, but also to detect imperfections which might otherwise escape notice.

In order to make perfect uniformity in the colour, the roller-boy should keep his ink well brayered out with the small roller, in proper quantities for the work in hand, and also should change his rollers well after taking ink, and at other times. The rollers are changed by moving the roller-handle slowly to the right and left, while the crank is being turned briskly with the left hand.

Torn or strained sheets met with in the course of work are thrown out and placed under the bank. Creases and wrinkles will frequently appear in the sheets when the paper has been carelessly wet: these should be carefully removed by smoothing them out with the back of the nails of the right hand.

If the frame of the tympan rub against the platen, it will inevitably cause a slur or mackle: this can easily be remedied by moving the tympan so as to clear the platen. The joints or hinges of the tympan should be kept well screwed up, or slurring will be the consequence. When the thumb-piece of the frisket is too long, it always produces a slur: this can be prevented by filing off a part of it. Loose tympans will at all times slur the work, and great care must therefore be observed in drawing them perfectly tight. The paper drying at the edges will also cause a slur: this may be remedied by wetting the edges frequently with a sponge.

Slurring and mackling will sometimes happen from other causes: it will be well in such cases to paste corks on the frisket, or to tie as many cords as possible across it, to keep the sheet close to the tympan.

The pressman should make the boy roll slowly, or the rollers will be apt to jump, and cause a *friar*. To prevent the rollers from jumping or bounding, bridges or springs made of thin steel, to reach across the gutters, may be used: these springs should taper off at the ends, and have an oblong hole in each end, through which they may be tacked to the gutter-sticks. In very open forms, it may be necessary to put bearers or pieces of reglet where the blank pages occur at the end of the form, to prevent one end of the roller from falling down and leaving a friar at the opposite end. This difficulty may, in a great measure, be obviated by imposing the form in such a manner as to bring the blank pages in the centre. This mode should always be adopted for title-pages and other light matter.

Before the pressman leaves his work, he covers the heap of paper by first turning down a sheet like a token-sheet, to show where he left off, and then putting a quantity of the worked-off sheets on it, and a paper-board if convenient. Laying the blanket on the heap after leaving off work is a bad custom. If the paper be rather dry, it will be well to put wet wrappers on it, after damping the edges well. If the form be clean, he puts a sheet of waste paper between the tympan and frisket, and lays them down on the form; if it be dirty, it must be rubbed over with clean ley, and several waste sheets pulled on it, as before directed, to suck the dirty ley out of the cavities of the letter. On his return to work in the morning, he takes care to wet the tympan, provided the type be worn. If there should be any pages in the form particularly open, the parts of the tympan where they fall must not be wetted.

THE LEY-TROUGH.

THE form being worked off, it is the pressman's duty to wash it clean from every particle of ink, not only for the cleanly working and well standing of the letter in the subsequent composing, but to save his own time in making ready when the same letter gets to press again. Many an hour is lost from not bestowing a minute or two in thoroughly cleansing and rinsing the form.

For this purpose, printing-offices are provided with a ley-trough, suspended on a cross-frame, and swinging by iron ears fixed somewhat out of the precise centre, so that the gravity of the trough will cause it to fall in a slanting position forward. This trough is lined with lead, the top front edge being guarded from the pitching of the forms by a plate of iron. The form having been placed in the trough, on its side, the pressman takes hold of the rim of the chase by the hook, or instrument for that purpose, and, laying it gently down, pours the ley upon it, and sluices it by swinging the trough on its pivots two or three times to and fro; then, taking the ley-brush, he applies it to the whole form, type, furniture, and chase; the ley is then let out into a receptacle, and the form well rinsed with clean water, by swinging the trough as before; the form is then lifted out, and consigned to the care of the compositor.

The ley is made of pot or pearl ash, or, what is better, of concentrated ley. A large earthen jar is usually chosen for the purpose; a sufficient quantity of ash or concentrated ley is added to the water to make it bite the tongue sharply in tasting.

The ley-brush is made large, the hairs close, fine, and long, in order not to injure the type, while sufficient force is applied to search every interstice in the letter where the ink can have insinuated itself.

PUTTING UP AN ADAMS PRESS.

FIRST, put one of the side pieces of the frame in an upright position, with a block under each of its ends two or three inches deep. Then put to its place against this side piece the large end piece of the frame, and secure the two firmly together by means of the bolts belonging to them.

Next, put to its place against said side piece, the vibrating frame on which rest the springs which help to sustain the bed and form of type, a journal at one corner of said vibrating frame being placed in a bearing provided to receive it on the lower part of said side piece. Then put one end of the bottom bar, or "winter,"—a heavy bar which sustains the force of the impression,—into an oblong square opening in the lower part of said side piece, that side of the bar on which provision is made to receive the toggle joints being toward the large end piece of the press-frame.

Next, put to its place against said side piece, and secure with bolts, the light end piece of the press-frame; and then insert one end of a small rod which serves as a stay to the upper rails of the frame, into a hole made to receive it, not far from midway of said upper rails.

Next, put up the other side piece of the frame, with blocks under its ends, as in the case of the first piece; put the bolts by which it is secured into their places, and, after levelling and squaring the frame, turn all the bolts and nuts up hard, so as to secure the frame in all parts with entire firmness. The bed must now be put into its place, the projections on ends being made to enter grooves formed on the side pieces of the press-frame, to guide it in its motions up and down.

Next, put the toggle joints into their places, their lower parts being jointed to the bottom bar, and their top ends to the under side of the bed, and the caps and screws which hold them to their places being put firmly together. The nature of the parts and the marks will indicate how this must be done. The pieces which serve somewhat as guides to keep the bed level as it rises and falls, should next be put on and bolted to their places under and at the ends of the bed. The impression cam (goose neck) should next be connected with the "bottom bar," and the bed so placed as to bring the roller in the middle joint of the toggle joints, in such position as to allow it to enter the mouth of the impression cam when said cam shall be raised up and pressed forward. Next, put the cams which operate the bellows and fly for taking off the sheets on the large gear-wheel shaft, the gear being on it, and then put the shaft into its bearings on the large end piece of the frame.

Next, put the fly-wheel shaft with the pinion on it into its bearings, *taking care to have the pinion and large gear-wheel go together, as marked.*

Next, put into its place the long shaft which has a bevel gear on each end of it to drive the distribution cylinder, and secure the box that supports said shaft. This shaft goes outside of the frame, and the bevel-wheel on one end is made to gear into another bevel-wheel, which is next to be put on the fly-wheel shaft and forced up to a shoulder on said shaft. There is a cover to go over one pair of these gear-wheels.

Next, put the fly-wheel on to its shaft, and force it up close to the bevel-wheel.

Next, attach the large connecting rod, or "pitman," to the large part of the crank-pin on the large gear-wheel, and also to the impression cam; and then, by turning the fly-wheel, raise up the bed and put the springs which help to sustain its weight into their places. To do this, after the bed has been raised up, first raise the vibrating frame on which the springs rest as high, on blocks, as will admit of the springs being pressed into their places.

The bed may then be drawn down by turning the fly-wheel until the rods on which the springs slide enter holes or boxes in the said vibrating frame provided to receive them, and until the bed shall be near enough to the vibrating frame to allow the frame to be connected with the bed by means of two flat connecting rods provided for that purpose.

The platen should next be put on in its place, and care must be had to put the shafts and wheels which sustain it on the top rails of the frame according to their *marks;* but, before these are put on, the small brass pieces, on which hang the rollers that carry the travelling tympan-sheet, should be connected in their places on the corners of the platen. Next, put to the ends of the platen the small pieces provided with thumb-screws, to prevent the impression bolts from turning while the press is in operation.

Now, screw the impression bolts into the ends of the bottom bar, first having put between said bolts and the frame certain small blocks of iron provided to keep them at a proper distance from the frame. In order to bring the work square, the impression bolts must be turned into the bottom bar, until certain small marks which are made upon the threads of their screws shall come even with the top surfaces of the ends of said bar, and so that the hole marked No. 1 on the head of each bolt shall come opposite to the points of one of the before-mentioned thumb-screws.

The impression bolts will now be in their proper relative positions, and will rest on set screws in projections on the lower parts of the frame, which set screws are adjusted to their proper height before the press leaves the manufactory, and do not afterward need to be altered. The impression bolts are kept in a perpendicular position by means of clamps, which are bolted on over them to the sides of the frame; and, whenever said bolts are turned for the purpose of increasing or lessening the impression, care must be had that one bolt be turned as much as the other. The holes in their heads are numbered for this purpose.

A cross piece which rests near to the bottom bar, on the ledges that project from the lower part of the frame, is next to be put into its place.

This cross piece serves to support one end of the rocker shaft which is used to disengage the main connecting rod, so as to prevent an impression being given, whenever it may be necessary to do so. It likewise supports a C spring, on which the impression cam falls, and an oblique stud, which the toggle joints bear against when in their place of rest.

In order to get the rocker shaft for preventing impression into its place, disconnect the large connecting rod from the impression cam, and let the cam turn down on to the aforesaid cross piece, and then put the shaft into its bearings.

The impression cam is now to be lifted up and again attached to the connecting rod, and then the bed may be raised up by turning the fly-wheel; after which, said C spring and oblique stud may be bolted to their places on said cross piece, and the bolts which hold the cross piece to the frame may be put in and screwed up firmly.

Next, put into bearings on the lower rails of the frame the rocker shaft which has arms for moving the frisket carriage; and also the shaft which has the vibrating cam (duck's bill) on it, for moving the arms that move the frisket carriage. Now, a stud that carries the friction rollers is to be put into slots in the arms that move the frisket carriage, said friction rollers being made to come within the oblong crooked openings in the last-named cams. There are cap-pieces which fit the journals of the two last-named shafts, and which are now to be put on to their places within the journal boxes. And there are cap-pieces with a set screw in the middle of each, which are to be bolted on firmly over the first-named cap-pieces, the set screws having first been turned back, so that when the cap-pieces are drawn down hard by the screws on to their seats, the journals shall not be bound in their boxes. After the outer cap-pieces have been bolted on as aforesaid, the set screws may be turned in until the under cap-pieces are secured firmly over the journals, but not so closely as to prevent said journals from moving freely in their boxes.

After the set screws are properly adjusted, secure them by means of the tightening nuts.

Next, put the frisket carriage in its place, and connect it with the arms that move it. Then put the four arms on to the small rocker shaft which operates the bellows that causes the sheets to rise from the frisket, and put said shaft into bearings which are located on the inside of the frame near to the fly-wheel shaft. Two of the aforesaid arms are made to bear against the under side of the bellows, so as to lift it; another of them is used to hang weight upon; and the fourth has a roller in its end, which lies upon a cam which operates it, and causes the weight to rise until the proper moment, when, by means of a sudden falling off in the cam, the said weight is allowed to fall, and give a sudden puff of wind from the bellows. Now, set the cam which operates the bellows close up to the shoulder on the main shaft, and so that the marks on the cam and shaft shall coincide with each other; then, by means of the set screw, secure the cam firmly. Next, put the bellows into its place: it is suspended by four bars, two of which rest on a stay rod that crosses the frame near the bed, and the others are bolted on to the under side of the top rail of the large end piece of the frame.

Next, put the pointing board into its place. This is supported at its upper edge by pivots, which rest in holes in the top ends of two small pillars; and these pillars (or studs) should first be put on to the pivots, and then the shanks of said pillars, both at the same time, must be put into the holes provided to receive them, in the top of the press-frame. Next, put the frame that contains a cluster of mahogany rollers into its place. This goes under the pointing-board, and is supported at two of its corners by pivots which rest in holes made for the purpose in the sides of the pillars that support the pointing-board. The other corners of this frame, as well as the two lower corners of the pointing-board, rest upon the top edges of the press-frame, near the platen. Said roller-frame must be put in with the vibrating roller uppermost. Next, put the pieces on to the press-frame that support the rocker shaft that carries the fly for throwing off the sheets; put said rocker shaft into its place, the spiral spring which turns it in one direction being at the same time put on to the shaft, and then put the fly-frame on to the said rocker shaft. Next, put the board that covers the cog-wheels into its place, and upon this board put the small pillars, or studs, and bar of wood upon which the fingers of the fly rest when receiving the sheets. Next, put into its place upon the end of the press-frame to which the fly is attached, the small shaft on which are several shives, or band pulleys. The journals of this shaft run in pivot-holes made in the top ends of the small brass studs, which stand in holes made in the top side of the frame to receive them. Both studs must first be put on to the journals, and then the two at the same time put into their places.

Next, put into its place on the end of the press-frame, a

rocker shaft that has two arms upon it, one of said arms having a roller in its end for a cam to operate against, and the other arm being provided with teeth on its outer end, which gear into a toothed pinion on the fly rocker shaft. A cam, before mentioned, on the main shaft, operates this rocker shaft and the fly. This is raised and lowered by means of a lever which passes along by the front side of the machine, between it and the person who lays on the sheets to be printed. Next, bolt on to the front side of the press a piece of wood which has a series of holes in it, by means of which the said lever by which the table is moved up and down is held in such position as may be desirable.

Next, put on to the inside of the press-frame a bent lever which has a ketch at one end, by means of which the fly is held down when no sheet is received upon it. The lever is operated by the foot of the pointer. A spring by which the motion of the fly is modified is next put on to one end of the pieces which support the journals of the fly rocker shaft. Next, at the bed end of the press, bolt to the press-frame the two brackets which form slides upon which to support the fountain; and then put into their places the levers which support the composition roller that carries the ink from the fountain to the distribution cylinder.

Next, put on to the inside of the press-frame, near to the end, the two hangers that support the distribution cylinder. Upon these hangers are small brass boxes in which run the journals of a vibrating cylinder, to distribute the ink laterally, and also the journals of two composition distribution rollers. Next, put into its place the said vibrating cylinder, the bores to receive the journals of which are situated in the lowest part of said hangers.

Next, bolt on two small pieces which have a friction roller in each to operate against the ends of the vibrating cylinder, to cause it to move to and fro in the direction of its length.

Next, attach to the inside of the press-frame, over the place for the distribution cylinder, a frame which has in it two wooden distribution rollers. These rollers are connected to each other by toothed wheels on their ends. This roller-frame is hung to the press-frame by means of a bolt at each of two of its corners, and the other two corners are supported on brackets, one of which projects inward from each side of the press-frame near to the bed. Next, put in two thumb-screws which serve to hold the distribution-roller frame down in its place: these screws are put through the sides of the press-frame from the outside near the top, and not far from the end.

Next, hang the large distributing cylinder.

Next, put on in front of the distribution cylinder the frame-work in which runs a carriage carrying a short composition roller, which is made to travel from end to end of the distribution cylinder, to and fro, by being pressed obliquely against it.

This carriage runs upon a rod supported by hangers from the two corners of the press-frame; and it may be taken out at any time by first taking out the rod on which it travels.

Next, put on to the inside of the press-frame, near its upper part, and near the edge of the platen toward the distribution cylinder, the wedge-like "click" which serves to push back the slide of the nipper-frame, to relieve the grip of the nippers.

Then put the nipper-frame into its place on the frisket carriage, and bolt on to the insides of the upper rails of the press-frame the two cams which raise the nippers and nipper-frame when advancing toward the pointing board to take the sheet which is to be printed. Next, put on to the pointing board the small bent lever that operates the "points," the end of the lever that connects with the point levers being put between the point levers and the sheet-iron plate of the pointing board. The other end of said lever, when the pointing board is in its place for work, passes down by the side of a spring on the inside of the press-frame, which serves to let down and hold up, alternately, the points at the proper periods in the operation of the machine.

The pointing board should always be lifted and lowered down, when raised up for any purpose, by taking hold of the side of the board next to the person who points the sheets; otherwise the lower end of said bent lever will not come into its proper place.

Next, screw a brass hook on to the under side of the frame of the pointing board. This serves to hold up both the pointing board and the cluster of mahogany rollers in nearly a perpendicular position, by being hooked on to the corner of the roller-frame. These parts are fastened up in this way in order that the platen may be drawn off from over the bed, to put on forms, and for other purposes.

Strong cotton tape bands may now be put on to the rollers

and shives, upon which to carry the printed sheets from the frisket on to the fly. One set of these bands pass around the roller nearest to the platen, and also around the roller that is so hung as to be capable of vibration. Said bands should be made sufficiently short to bring the arms on which the vibrating roller hangs into a perpendicular position.

The other set of bands pass around the second roller from the platen, over the third roller from the platen, under the vibrating roller, and around the shive pulleys between the fingers of the fly.

To drive the mahogany rollers and the shive pulleys, a flat band runs from a pulley on the fly-wheel shaft over a pulley on the shive-pulley shaft; and then a small round leather band is put around a grooved pulley on the shive-pulley shaft, crossed and carried around the vibrating roller. The drawer, or frame which answers to the tympan-frame of a hand-press, is best covered with parchment or India-rubber cloth. The latter is now preferred by printers when it is good, and of *even softness*. If it is not of even softness, an even impression cannot be produced with it. Fine linen will answer, but it will not admit of so fine an impression as either of the other substances.

It will be obvious to the person who sets the press up, how the drawer goes on to the face of the platen. One edge of it is held up by lapping a little on to the edge of the platen, and the other by slide bolts at the corners. The same kind of blanketing may be used as in the hand-press; but, when the drawer is covered with India-rubber cloth, no blanketing is needed, except on a very heavy form. The tympan-sheet should consist of a very long sheet of strong smooth paper, the width being equal to the length of the platen; or else it should consist of a web of the finest kind of cambric. This is drawn under the platen, and its ends attached to rollers, one of which rollers lies in bearings at one side of the platen, and another at the other side of the platen. By means of these rollers, the tympan-sheet is moved from place to place, as it becomes soiled by use, until it has been used as much as proper throughout its length. It will bear using several times over. If it consists of cambric, it may be washed. For some kinds of printing, it is best that the tympan-sheet should be moved periodically the whole width of the form at a time; and in this case the rollers for the purpose must be turned by hand. But generally it is more convenient for the rollers to be turned by the machinery in such manner as to move the tympan-sheet a very short distance at each impression. To effect this movement, there is a small lever suspended from the end of one of the spindles of one of the tympan-sheet rollers, and this lever carries a "click" that operates on a ratchet wheel on the end of the roller spindle, to turn it. The lever is operated by a small crank screwed on to the outer end of the main shaft, the crank operating on the lever by means of a rod connected to the crank by one end, and operating on the lever by a hook at the other end. The hook end of the rod is supported by a piece screwed into the side of the frame for the purpose, the rod sliding in a bearing formed in said piece. Very fine flannel is used by some for tympan-sheets. The table that the "heap" is placed upon stands upon the platen, and its top is adjustable.*

MAKING READY ON CYLINDER PRESSES.

Make clean the bed of the press and the impression segment of the cylinder. Adjust the bearers a trifle above ordinary type height. See that the impression screws have an even bearing on the journals, and that the cylinder fairly meets the bearers. Select a suitable tympan or impression surface.

The tympan may be the India-rubber cloth which is furnished with the press, a thick woollen lapping cloth or blanket, several sheets of thick calendered printing paper, or one or more smooth and hard pasteboards. Each of these materials have merits not to be found in any other. Upon the proper selection of the tympan the presswork in great measure depends, and the pressman should be thus guided in making choice.

A pasteboard tympan is most suitable for wood-cuts, for perfectly new type, and for the best kinds of presswork. It is not suitable for miscellaneous work, nor for heavy forms, nor mixed old and new type. If the overlaying is properly executed, a pasteboard tympan will enable the pressman to show a sharper edge and a more delicate impression of the type than can be possible with any other, and it will wear the type less than any other. But it will require a very tedious and careful making ready, or it will prove very destructive to type.

* R. Hoe & Co.'s Catalogue.

A woollen blanket is best adapted for old stereotype plates, for very old type which has been rounded on the edges, for posters with large wood type, and for all common work which requires a clear but dull impression. For such work a woollen blanket will enable the pressman to make ready a form more quickly than with any other material; but it is injurious to new type, and incapable of producing a fine and sharp impression.

Thick paper is much used for book-work. It also answers well for script circulars and leaded forms. It will not answer so well for mixed old and new type, nor for table-work with unequal heights of brass rule, nor for mixed large and small type. It will prove most serviceable for the average of light and fine press-work.

The India-rubber cloth combines many good qualities not found in other tympans: it has something of the density of the pasteboard, the hardness and evenness of paper, and the flexibility of the blanket, combined with an elasticity peculiarly its own. It will compass a greater variety of work than any other: posters, script circulars, news and book forms, stereotype plates, and old or new type, can all be well printed with an India-rubber blanket. When it is intended to make one tympan answer for all kinds of work, the India-rubber blanket will be found decidedly superior to all others; but when very extra press-work is wanted, the tympan must be specially adapted to the form of type.

There are forms for which none of these tympans are specially suitable. For such cases careful pressmen combine two or more together,—as Welsh flannel over rubber, or thin rubber over pasteboard or under paper. These, however, are exceptional cases, and are only thus combined when very good press-work is wanted from imperfect materials. Careful observation of the *quality* of the impression given by each style of tympan will teach a pressman how to combine to the best advantage. As it requires experience and discrimination, an arbitrary rule cannot be given.

Whatever may be the material selected, the tympan must be stretched very tightly over the cylinder. All labour in overlaying is but thrown away if this is not carefully attended to. A rubber or woollen blanket can be secured at one end of the cylinder by small hooks projecting inward, while it may be laced tightly with saddler's thread at the other end; or, by sewing on that end of the blanket a piece of canvas, it may be wound tightly around the reel, and kept secure by the pawl and ratchet.

But paper and pasteboard require a different process,—viz.: Take a piece of Nonpareil cherry reglet of the full length of the cylinder. Trim down the paper or pasteboard to the width of the bed between the bearers, but leave it a little longer than the impression segment of the cylinder. Then crease the pasteboard at a uniform distance of half an inch from the narrower end, and lay this creased part on the flat edge of the impression segment of the cylinder, under the grippers. Put the reglet over this, and bring down the clamps hard on the reglet, so as to bind all securely. When this is done, a thin web of muslin may be stretched over the whole, in the same way in which a blanket is laid on, and rolled up tightly, which will prevent any slipping of the board or of the overlays pasted on it.

A large poster, or newspaper form, or any large form with old type, will require a soft roller with much suction. Book-work, wood-cuts, or fine job-work, will require a harder roller, with very smooth, elastic, and clinging surface. Coloured inks are best printed with a still harder roller and with much less suction. All rollers should be perfectly clean, and free from cracks or holes. The suitableness of these rollers cannot well be explained by words: such a knowledge will be best acquired by observation and experience. It may, however, be necessary to state that one roller will not answer for all styles of press-work: the quality of the work, the size and wear of the type, and the speed of the press, must control the pressman in his choice.

Posters, with large wood type, require a semi-fluid ink, but not surcharged with oil. Ordinary news-work requires a better grade, more tachy, and finely ground. Good book-work should have a stiffer-bodied ink, soft, smooth, and with little oil. Job ink, which is made expressly for presswork on dry paper, should be used only for such work. Book and job inks are not convertible: an ink for wet paper will not work well on dry paper, and *vice versâ*. Very fine press-work—such as wood-cuts, or letter-press upon enamelled paper—calls for an ink impalpably fine, very stiff, of brilliant colour, and nearly or absolutely free from oil.

Every job-office should keep four grades of ink,—news,

book, job, and wood-cut. They can be compounded (if no ink-manufacturer is near) with each other, or reduced with varnish to suit any form. Good press-work is impossible without good inks.

Charge the ink fountain with the ink selected, and keep it well covered, to protect it from paper dust. Turn down the screws, and cut off all the ink evenly. When the form is ready, turn on the ink cautiously, and wait for ten or twelve impressions before again altering the screws. For small forms and short numbers of any piece of press-work in coloured ink or extra ink, a fountain is not necessary. The ink may be applied with a brayer or palette-knife.

The adjustment of the margin is the next process. Although type can be printed from any quarter of the bed, it will be found most judicious to lay all forms close to the back part of the bed, and equidistant between the bearers. This will secure a good impression, give a fair average margin to every form, and allow the full use of the bed for a large form, without resetting the cylinder. The bed and cylinder travel together, and the grippers, which bring down the sheet to the form, should barely lap over the back part of the bed. So long as the toothed cylinder wheel, and the short toothed rack on the side of the bed, remain undisturbed, the grippers will always pass over the bed in exactly the same place. When the grippers are in this position (slightly lapping over the back of the bed), take measurement of the distance between the back edge of the bed and the point of one of the nearest grippers. With a piece of reglet cut a gauge exactly corresponding to this measurement. Let no form be laid upon the press until the distance between the type and the edge of the chase tallies with the gauge. This will prevent the grippers from closing on the form and crushing the type. If the chase will not admit of so wide a margin, or if an extra margin is wanted on the sheet, put a piece of furniture of the extra width behind the chase: the margin can thus be increased or diminished at pleasure.

A book form may be locked up in a chase so large, with the type so far from the edge of the chase, that the grippers will bring down the sheet in such a position that it will be printed with the margin all on one side. To remedy this, the cylinder must be reset. Proceed thus. Remove the screw and washer, and draw the intermediate wheel out of gear, loosen screws in the gauge rack, then turn the cylinder to the point required, connect the intermediate wheel, adjust the gauge rack, and screw up tight.

The press having been adjusted, next examine the form to be printed. Not only see that it is gauged correctly, but also see that it is not locked up too tightly,—that chase, quoins, letter, and furniture are all level, and lie flat upon the bed. If the form springs, the quoins must be slackened; if this loosens the type too much, the justification should be amended. Make clean the type by rubbing it over with a dry brush. The rollers are often made foul and the colour of the ink changed by dust and particles of dirt clinging to the type.

Fasten the form so securely on the bed that it will not be moved by the action of the cylinder or the rollers. Take a proof on its own paper, using very little ink. Proceed to adjust the drop guides so as to bring the sheet exactly in the right position. Push out the iron tongues at the edge of the feed-board, and at equal distances from each other, so that they will equally sustain the paper. Slide the drop guides along the rod until they fall squarely over the tongues. Set the side guide so that it will give a true margin in length to the sheet to be printed. Adjust the grippers so that they will seize the sheet at proper intervals, making the margin exactly even by lengthening or shortening the drop guides. Then take a clean proof *on its own paper*, exactly in the right position, before making ready, when it may be shown to the reader. It frequently happens that an error in the margin, or an imperfection in the register, is thus noticed; and its timely discovery and correction before overlaying will save much time and trouble. A readable proof may be taken before overlaying, by running through a sheet or two of proof paper. *Make register*, if it is a book form, *before underlaying.*

When every thing has been found correct, then proceed to regulate the impression. If the type is fair, the proof should show a decently uniform impression; but if the form is large, or if it contains old and new or large and small type, then the proof will show an uneven impression. To rectify this inequality, pressmen use many expedients.

1. By lowering the bearers and putting on more impression. This, of itself, is a very poor way; for it wears down new type in order to show the face of the old, and invariably produces thick and coarse presswork.

2. By raising the low type to a proper height with thicknesses of paper under them, which is called underlaying.

3. By giving additional thickness to the tympan over every part of the form which shows a weak impression, which is called overlaying.

It is very rare that any one of these modes will prove sufficient: all should be used in conjunction. When the larger part of the proof-sheet shows a weak impression, almost approaching illegibility, then more impression should be added. When one side of the proof-sheet shows a weak impression, while that on the other side is full and clear, then more impression should be given to the paler side. The impression should be made decently uniform before any attempt at overlaying or underlaying. But the bearers should follow the impression screws, both being raised and lowered together, in order to secure the type from the unimpeded force of the impression cylinder. Not only should the bearers be of even height, but the cylinder shaft should always revolve on a true level. If the impression screws are carelessly used, and the bearers are rashly raised and lowered, this even bearing will soon be lost; the difficulty of obtaining a good impression will be much increased, and the press will receive a serious injury. For the same reason, the bearers should never be packed (by the addition of cards, as is usual on a hand-press), for it strains the cylinder and all its bearings with an irregular resistance. The bearers should be tampered with even less than the impression screws. When the impression screws are so set that the cylinder gives a fair uniform impression, they have done all that can be expected, and nothing more should be attempted by them. Sometimes the proof may show that a cut, or a line of type, or a set of brass rules, are higher than any other material in the form. But the impression should be set regardless of this: it will be found quicker and neater to reduce the impression on one or two such high lines, by cutting *out* the tympan-sheet over them, than it would be to underlay and bring up all other types to such an irregular height. Pitch the impression so that it will face the larger portion of the type, and make the less conform to the greater. Those parts which are high must be cut out of the tympan, and those which are low should be raised by underlays, and all inequalities smoothed by overlays.

When any part of the form is very low, it will not answer

to attempt facing it with overlays: it must be brought up to meet the inking rollers as well as the impression cylinder. When the proof shows low type, cut out the impression of it, raise the form, and paste it over the feet of the letter. If some types are high and some are low, make proper distinction, and carefully avoid increasing the height of any type or rule which seems to have a full impression. Pursue the same course when a marked depression appears in the centre, or a dwindling impression at the edges. Cut out that section which is light, and affix it to the defective part. If the impression dwindles in any part, the underlays must be cut of irregular thickness to suit the tapering off of the impression. Cut out an underlay from the edge where the impression begins to fade; then cut another of smaller size where it is utterly illegible; paste one over the other, laying them carefully in their proper positions, and then paste them all on the bottom of the form, where it is needed, taking care to lay the smallest underlay nearest the bed. This will restore the type to a proper level, and the next proof should show a uniform impression. The same plan will answer for a low corner. Use as little paste as possible, thin and free from lumps. Be careful that the underlays are laid on smoothly, without fold or wrinkle. Cut all underlays from a proof; for the proof serves as a guide both in cutting and in affixing to the form.

Underlaying should not be done to any great extent upon a cylinder press. It is a valuable means of bringing up an old line of type, a hollow or a low corner. The underlays of any *type* form should not constitute more than one-fourth of the surface; if more than this is attempted, they will rarely ever fail to work up the quadrates and furniture. The action of the quickly moving cylinder upon a form of type underlaid with yielding paper, must create a spring and a rocking of all the materials in the chase.

Of all materials, old stereotype plates need underlays most, as they are usually very irregular in height. Thin card or pasteboard will be found preferable to paper for the underlaying of plates secured on wood bodies. When the plates are on patent blocks, always underlay between the plate and the block. Always cut the underlay for a plate less in size than the faint impression would seem to require. This will allow for the spring of the plate. If it is cut of full size, the next impression will disappoint the pressman, by being much harder at the edges than he intended. Never attempt to build up a type form to a proper impression entirely or chiefly by underlaying.

Underlays should be put under all large and bold-faced types when used with much smaller types, so as to raise them above the level of the others. This is needed to give it closer rolling, extra supply of ink, and that extra force of impression to transfer the ink to paper which all large types require.

When the type has been so levelled by underlays that all parts receive proper bearing from the inking rollers, and when the cylinder has a corresponding even impression, then overlaying may be commenced. For ordinary news, posters, or job-work, overlaying may be entirely unnecessary; the tapes and fly may be set, and the printing of the form may proceed without further delay. But fine presswork cannot be done without overlays. Underlays are chiefly valuable for securing an even impression; while overlays are indispensable for giving delicacy and finish to that impression.

To overlay a form properly, the tympan should be covered with a sheet of thin, smooth, and hard paper, stretched tightly. Then take a pale impression on the tympan-sheet, and also run through the press two or three proofs on thin and hard paper. Examine the proofs carefully on both face and back. If any brass rules or letters appear too high, cut them out of the tympan-sheet in one or two thicknesses, as their varying height may require. Go over the whole proof, examining every line carefully, and, by cutting out, reduce the impression of all projecting letters to a uniform standard. For this, as for all other work on overlays, use a very sharp knife with a thin point, and cut on a smooth surface, so that there will be no dragged or torn edge to the cut.

The next step should be to raise the impression of those parts of the form when the type appears dull or weak. Cut out carefully, and paste the overlays over the tympan smoothly. Overlays are worse than useless if they are not laid on firmly and smoothly, as the slightest bagginess will cause slur or mackle. If, by accident, the tympan-sheets or overlays should bag or wrinkle, tear them off, and commence anew.

Cut out and overlay the more prominent parts first; then try another impression, and from that cut new overlays for minor defects. Thus proceed until a perfectly smooth and even impression is obtained.

With common work it will be sufficient to cut overlays in masses, as pages or parts of pages; but with fine work every line and letter needs examination, and letters and parts of single letters are often overlaid by careful pressmen. When the pressman is expert at making ready, it is not necessary that he should take a new impression with every successive set of overlays. Many pressmen take a dozen proofs of a form on different styles of paper, and proceed to cut out and overlay on one of the proofs, and finally paste this proof on the tympan. But this boldness and precision can be acquired only by long practice. It is better for the young pressman to feel his way step by step.

The Impression.—A diversity of opinion exists among good printers as to the proper force of the impression: by some a heavy and solid indention of the paper is considered necessary; while others insist that an impression which does not indent the paper is preferable. But the indention of the paper is no test of the force of the impression. A light impression against a woollen blanket will show more forcibly than a strong impression against a paper or a pasteboard tympan.

Type is worn out not so much by the *direct* impression of the platen or cylinder on the flat face of a form, as by a grinding or rounding impression on the edges of the type, caused by the forcing of the tympan between the lines and around the corners of every letter. Every fount of worn-out type, whether from cylinder or platen press, has suffered less from a reduction in height than from a rounding of the edges. When the type is new and the tympan hard and smooth, the impression can be made so flat that the type will not round at the edges, and the impression will not show on the paper. But this cannot be done with old type or with a soft tympan: the impression must be regulated to suit the tympan. On fine work, a rounding impression should be avoided, as it not only destroys type, but also thickens the hair-line and wears off the ceriphs.

It is not sufficient that the paper should barely meet the type: there must be sufficient force in the impression to transfer the ink from type to paper. If there is not sufficient impression, it will be necessary to carry much ink on the rollers; and this produces two evils: the type is clogged with ink, and the form becomes foul; too much ink is transferred to the paper, which smears and sets off for want of force sufficient to impress it in the paper. Distinction must be made between a *light* and *weak* impression and a *firm* and *even* impression. The latter should be secured even if the paper is indented; though that is not always necessary. But a form of old type, a poster, or other solid form, must have a *heavy* impression, or else a very tedious and careful making-ready.

To set New Tapes.—Pass the tape around and close to the cylinder. Lap it over one of the tape pulleys, and then pass it around the small guide pulley on the shaft above. To increase its tightness, throw up the guide pulley from the shaft, and set the binding screw more tightly. All these pulleys are movable on their shafts, and distance between them may be altered at pleasure. Let the tapes rest upon the outer margin of the sheet, and see that the overlays on the tympan over which the tapes pass are of equal thickness; if not of equal thickness, the sheet will wrinkle.

To set the Fly.—Run through a sheet of the paper to be printed, and let it run down the fly so far that it is barely held by the fly pulleys. Then set the cam which works the fly, so that its point just clears the small friction roller on the shaft, and it will throw down the sheet correctly. Tighten the spring according to the size of the sheet, and set the spring crank so that it will prevent the fly from striking too hard on the table.

It will be seen that good presswork does not depend *entirely* upon the press, nor yet upon the workman or the materials. Nor will a superiority in any one point compensate for a deficiency in another: the newest type will suffer from a poor roller, and the most careful making-ready will be of no avail if poor ink is used. It is necessary that *all* the materials should be of the best kind,—that they should be well adapted to each other, and fitly used. Although a good workman can do much with insufficient materials, there are cases where a neglect to comply with one condition is equal to a neglect of all.*

FINE HAND-PRESSWORK.

FINE presswork is the art of printing perfect impressions from the surface of type or engravings in relief: that is, the subject transferred to paper should be an impression from the

* R. Hoe & Co.'s Catalogue.

surface, and the surface only, of the types or engraved lines, of such a tone as to produce all the effect of which the subject is capable, without either superfluity or deficiency of colour.

The press ought to be in the best condition; otherwise it will be impossible to get an equal impression without much trouble and loss of time. The joints of the tympan should not have any play, or the correctness of the register will be affected, and slurs and doubles be caused.

The parchments on the tympans should be thin, and of a uniform thickness, and stretched on the tympans so as not to be flaccid. On account of its thinness, smoothness, and uniformity, silk is probably preferable.

The face of the platen ought to be a true plane, and parallel to the press-stone, or table.

The advantage of having a good press is unavailing for the production of fine work if the types are much worn; for it is impossible to produce a sharp, clear impression when the type is worn and the fine lines are rounded by much use. In consequence of this roundness of the letter, it is necessary to use a thick blanket in the tympan to bring up the type; thus producing a gross and irregular impression of more than the surface.

Ink for fine work should be characterized by the following peculiarities:—

Intenseness of colour.

Impalpability.

Covering the surface perfectly of the type or engraving.

Quitting the surface of the type or engraving when the paper is pressed on it, and adhering to the paper.

Not smearing after it is printed.

Complete retention of colour.

Ink ought to be reduced to an impalpable smoothness, either in a mill or on a stone with a muller. This is essential, as the process gives it the next quality,—that of completely yet very thinly covering the surface of the type or the lines of the engraving, and insuring an even and perfect appearance to the impression on the paper. Another important requisite is, that the ink shall not only cover the surface of the lines on the paper printed, but that it shall also quit the face of the type or engraving and leave it quite clean when the paper is impressed on it, and attach itself to the paper, so as to give a perfect impression of the subject represented, without the colour of the paper appearing through the ink; and that this peculiarity of quitting the type or engraving and becoming attached to the paper shall continue the same through any number of impressions, without any accumulation of ink on the surface printed from. After having obtained these results, and when the printing is as perfect as it can be made by workmanship, something more is requisite,—viz.: that the ink shall not smear on being slightly rubbed, and that it shall retain its colour and appearance without spreading at the edges or tinging the paper.

The rollers should be in good condition; otherwise the pressman may exert his skill in vain, with a great loss of time and waste of paper.

The quality of the paper is of great consequence in fine printing; but it is frequently overlooked by the printer's employers, who are apt to pay more attention to a showy appearance and a low price than to quality.

The best paper for printing on is that which is made of fine linen rags and moderately sized, without the use of acids in bleaching, and without being adulterated with cotton rags: this paper takes water kindly, is easily got into good condition, receives a good impression, is durable, preserves its colour, and does not act upon the ink.

The use of cotton rags, the introduction of gypsum into the manufacture of fine and other papers, the application of acids and bleaching powders to improve the colour and produce apparently good paper from an inferior staple,—these form the grand hinderances to the American printer in his efforts to equal or excel foreign productions. Hence it is that works printed in this country are less valued than those from the English press, which are printed on paper of a fine fabric, made mostly of linen rags, and sufficiently strong to bear a fine ink.

A pressman should, as a matter of course, be well acquainted with the entire routine of presswork; in addition to which, to form his judgment, he should examine the most splendid productions of the press, and study them as patterns of workmanship.

In making ready, it must be evident that, when a clear, sharp impression is wanted, the pressure should be on the surface only. Of course the tympan ought not to be very soft, neither should a woollen blanket be used: the most perfect impression will be obtained when fine thick paper alone is placed in the tympans; and even of this article but few thicknesses should be employed.

After an impression is printed, the pressman examines if it be uniform throughout; if it be,—which is very rarely the case,—he goes on with the work; if not, he proceeds to overlay, in order to produce regularity of pressure and of colour over the whole form. Wherever the impression is weak he pastes a bit of thin, smooth paper, of the size and shape of the imperfect part, on the tympan-sheet; he then pulls another impression, to examine the effect of his overlays, and continues to add to them where wanted, till the pressure of the platen is the same in every part and the impression is of a uniform shade of colour.

If the impression come off too strong in parts, or at the edges or corners of the pages or on the head-lines, it will be necessary to cut away the tympan-sheet in those parts, and, if that does not ease the pressure sufficiently, to cut away the same parts from one or more of the sheets that are within the tympans.

It is generally preferable to overlay on a sheet of stout smooth paper inside the tympan, particularly where the same press does the whole or great part of a work: this sheet is cut to fit the interior of the tympan, so as not to slip about, and has overlays pasted on it where wanted, to bring up the impression till it is very nearly equal. In all succeeding sheets it saves the pressman a great deal of time, as he will be certain that when he pulls a sheet of another form of the same work it will be nearly right, and he will only have to place thin overlays on occasional parts to make the impression perfect.

It is necessary, where short pages occur in a form, to have bearers to protect their bottom lines and the edges of the adjoining pages. These may be of double pica reglet pasted on the frisket, so as to bear on some part of the furniture or chase; but bearers made to the height of the types are better, when they can be used.

It happens occasionally that the tympan causes the paper to touch the form partially on being turned down, and occasions slurs. This may occur from the parchment being slack or the paper being thin and soft. To prevent this inconvenience, it is customary to roll up a piece of thick paper and paste it on the frisket adjoining the part. Many pressmen prefer pieces of cork cut to about the thickness of double pica, and pasted on the frisket.

In working the white paper, instead of pins stuck into the tympan, to prevent the paper slipping, a duck's bill (a tongue cut in a piece of stout paper) is frequently used: it is pasted to the tympan at the bottom of the tympan-sheet, and the tongue projects in front of it; indeed, the tympan-sheet appears to rest in it. The bottom of each sheet is placed behind this tongue, which supports it while the tympan is turned down.

The rollers should be kept clean, but should not be too moist, as this will prevent the ink from distributing equally over them, and from covering evenly the surface of the types or engraving; nor should they be too dry, as in that case they will not dispose of the ink smoothly enough to produce a fine impression, neither will they retain particles of dirt on their surface, but will part with them to the form, thus causing picks.

The ink ought to be rubbed out thinly and equally on the ink-block, so that when it is taken it may be diffused smoothly over the surface of the rollers. It is advisable to keep rubbing the ink out on the block with the brayer, and to distribute the roller almost constantly; the continual friction produces a small degree of warmth, which is of advantage, particularly in cold weather.

As uniformity of colour is requisite for beauty in printing, where the form is large the pressman should take ink for every impression: this may be thought troublesome, but it is advantageous in producing regularity of colour. It is unpleasant to see in a fine book two pages that face each other differing in colour,—the one a full black, surcharged with ink, the other deficient in quantity and of a gray colour; yet this must happen when, as is frequently the case, three or four sheets are printed with one taking of ink.

In fine books, particularly where the paper is large and heavy and the type large, set-off sheets are used to interleave the whole impression while working: these remain till the printed paper is taken down from the poles by the warehouseman. These set-off sheets are put in when the white paper is working, and moved from one heap to the other during the working of the reiteration. They prevent the ink from setting off from one sheet to another while they are newly printed, from the weight of the paper.

To secure uniformity of impression, the pull should be so adjusted in the first instance as to give a proper degree of pressure on the form when the bar is pulled home; then, checking the bar, it should be allowed to rest in that position during a short pause.

It will be perceived that, to produce presswork of a highly superior character, great expense and much time are required; and that it is requisite to have a good press in good condition; to have new types, or types whose faces are not rounded by wear; to have good rollers in good condition; that the ink should be strong, of a full black colour, that will not fade nor stain the paper, and ground so fine as to be impalpable; the paper should be of the best quality, made of linen rags, and not bleached by means of acids or bleaching powders, which have a tendency to decompose the ink; the rolling should be carefully and well done; the face of the type should be completely covered with ink, without any superfluity, so as to produce a full colour; and the pull should be so regulated as to have a slow and great pressure, and to pause at its maximum in order to fix the ink firmly upon the paper. These particulars observed, with nothing but paper in the tympans, perfect impressions of the face alone of the type will be obtained, and a splendid book will be produced in the best style of printing.*

PRINTING WOOD-CUTS.

A SINGLE block, when imposed in a large chase, may spring out of the chase while being inked, from the quantity of furniture about it. A good plan is to impose it in a job-chase, and to impose this chase in a larger one: this will cause it to lie flatter on the press and firmer in the rolling, as the small chase can be locked up tight in the large one, without having too much furniture, and the large one can be secured firmly on the press by quoins and the corner irons.

Before pulling the first impression, the workman should see that the surface of the cut is perfectly clear from particles of dirt, and that no pin or lump of paste is on the tympan. He ought then to pull very gently, or he may injure some of the fine lines of the engraving.

* Adapted from Savage.

Neither the pressure nor the impression of an engraving on wood should be uniformly equal: if it be, the effect intended to be produced by the artist will fail; and, instead of light, middle tint, and shade, an impression will be produced that possesses none of them in perfection: some parts will be too hard and black, while other parts will have neither pressure nor colour enough, nor any of the mildness of the middle tint, which ought to pervade a large portion of an engraving, and on which the eye reposes after viewing the strong lights and the deep shades.

To produce the desired effect, great nicety and patience are required in the pressman: a single thickness of thin India paper (the best for overlaying), with the edges scraped down, is frequently required over very small parts. The overlay should never be cut at the edges; but, even where great delicacy of shape is not required, it should be torn into the form wanted: this reduces the thickness of the edges, and causes the additional pressure to blend with the surrounding parts.

As particular parts of the impression will frequently come up too strong, and other parts too weak, it will be necessary to take out from between the tympans a thickness of paper and add an additional tympan-sheet, cutting away the parts that come off too hard, and scraping down the edges. Scraping away half the thickness of a tympan-sheet in small parts that require to be a little lightened will improve the impression. The light parts require little pressure; but the depths should be brought up so as to produce a full and firm impression.

If a block be hollow on the surface, underlaying the hollow part will bring up the impression better than overlaying it,—at least so nearly that only a thickness or two of paper will be needed as overlays. If a block be too low, it is better to raise it to the proper height by underlays than to use overlays; as the latter act in some measure as blankets, and are pressed into the interstices, rendering the lines thicker than in the engraving.

It will be necessary sometimes, when the surface of the block is very uneven, to tear away parts of the paper in the tympan, to equalize the impression where it is too hard.

The pressman will find it convenient to pull a few impressions, while he is making ready, on soiled or damaged India paper. Out of these he can cut overlays to the precise shape and size wanted, which are always necessary when great accuracy is required in overlaying particular portions. He should be provided with a sharp penknife and a pair of good small scissors. A fine sharp bodkin and a needle or two, to take out picks, are also needful; but he should be particularly careful so to use them that he do no injury. The best way is, to draw the bodkin or needle point cautiously in the direction of the lines.

Engravings in vignette form require great attention to keep the edges light and clear, and in general it is necessary to scrape away one or two thicknesses of paper in order to lighten the impression and keep it clean: the edges being irregular and straggling, they are likely to come off too hard. Bearers type-high placed beside the block will be found advantageous; if they cannot be used, pieces of reglet, pasted on the frisket in the usual way, and taking a bearing on the furniture, must be substituted; but the high bearer is to be preferred where it can be adopted. The bearers equalize the pressure on the surface of the engraving, and protect the edges from the severity of the pull, which is always injurious to the delicacy of the external lines. They also render the subject more manageable, by enabling the pressman to add to or diminish the pressure on particular parts, so as to produce the desired effect.

When great delicacy of impression is required in a vignette, it will be found beneficial, after the engraving is inked, to roll the extremities with a small roller without ink: this will not only take away any superfluity of ink, but will prevent picks, and give lightness and softness to the edges, particularly where the effect of distance is required.

If the extremities are engraved much lighter than the central parts, underlays should be pasted on the middle of the block, which will give a firmer impression to the central parts of the subject. It would save trouble and aid in getting a good impression if the block were engraved a little rounded on the face.

When highly finished engravings on wood are worked separately, woollen cloth, however fine, should never be used for blankets, as it causes too much impression; a sheet or two of hard smooth paper between the tympans is better; sometimes even a piece of glazed pasteboard is used inside the outer tympan. The parchments ought to be in good condition, stretched tight, of a smooth surface, thin, and of regular thickness, so as to enable the pressman to obtain an impression as nearly as possible from the surface only of the engraved lines.

The rollers must be kept in perfect order; and the pressman should be very particular in taking ink and inking the block. He ought to use the best ink that can be procured.

When a wood-cut left on the press all night has become warped, lay it on its face upon the imposing-stone, with a few thicknesses of damp paper underneath it, and place over it the flat side of a planer, with sufficient weight upon it; in the course of a few hours the block will be restored to its original flatness. This method is preferable to steeping the block in water; as the steeping swells the lines of the engraving, and, consequently, affects the impression. To preserve the original effect of the cut as it came from the hands of the artist, the block should never be wet with water; and, when it has been worked in a form with types, it should be taken out before the form is washed.

To prevent warping during the dinner-hour or the night, turn the tympan down upon the form, run the carriage in, and, pulling the bar-handle home, fasten it so that it will remain in this position during the interim.

However long boxwood may be kept in the log, it will always twist and warp when cut into slices for engraving, on account of fresh surfaces being exposed to the air. Large blocks may be restored to their flatness in the course of a night by laying them on a plane surface, with the hollow side downward, without any weight on them.

A fine engraving on wood should never be brushed over with ley: the best method is to wipe the ink off with a fine sponge damped with spirits of turpentine, and, if it get foul in working, clean it with a soft brush and spirits of turpentine; then wipe the surface dry and pull two or three impressions on dry waste paper. Spirits of turpentine take off the ink quicker, and affect the wood less, than any other article. The facility with which the block is again brought into a working state more than compensates for the trifling additional expense incurred.

When a few proofs only are wanted from a small engraving, good impressions may be obtained with little trouble on dry India paper, with about six thicknesses of the same sort of paper laid over it, and pulled without the tympan; if proofs are wanted from large ones, it will be found advantageous to put the India paper for a few minutes into a heap of damp paper.

To do full justice to an engraving, the pressman should get a good impression from the engraver and place it before him as a pattern, and then arrange the overlays, &c. till he produce a facsimile in effect. Better still is it for an unpractised hand to obtain the assistance of the artist at the press-side, to direct him in making ready the cut.*

CARD PRINTING

Has, since the introduction of enamelled or polished cards, made rapid strides toward perfection; the fine absorbing quality of the enamel, under proper management, producing the most beautiful results,—in many cases scarcely discernible from copper-plate. A card, to be well printed, requires the same treatment as a wood engraving (see p. 258), so far as making ready is concerned, and in working without blankets and using the finest ink. Having made a light impression on the tympan-sheet, place the pins so as to bring the impression as nearly as possible in the centre of the card, one pin at the lower side and two at the off side, taking care that the head of the pin does not come in contact with the types. The impression should be exceedingly light until properly regulated,—at no time more than is actually necessary to bring up the *face* of the type. Cards are now mostly printed on small card-machines, at the rate of one, two, and even ten thousand per hour. All cards should be printed dry.

GOLD PRINTING.

The types being made ready for press in the usual manner, the surface is covered in the ordinary way with gold size instead of ink, and the impression taken upon the paper. For a large job, remove only the back from a book of leaf-gold; for a small one, lay a straight-edge across the book, and cut it through, of the size required, with the point of a sharp penknife. This must be done before using the size. Slightly wet the end of the forefinger of the right hand, and, placing the thumb of that hand on the pile of gold, raise the edge of the

* Adapted from Savage.

paper with the forefinger sufficiently to dampen it with the moisture of that finger; then press the moistened edge of the paper on the gold, and it will adhere sufficiently to enable the fingers to lift gold and paper together and place it on the impression. Proceed thus until the size is entirely covered; gently pat the gold with the balls of the fingers, or any soft, pliable substance, until it is set; then, with a very soft hat-brush, remove the superfluous gold, when a clear and beautiful impression will appear. Its sharpness will depend on the judgment of the printer in applying the size to the type.

BRONZE PRINTING

Is used more extensively than gold printing, being attended with far less expense in the cost of the material. The method of printing is the same, except that, instead of laying on the gold-leaf, the impression is rubbed over with the bronze, by dipping a small block, covered with a short, fine fur, into the powder, and brushing off the superfluous bronze with a soft brush, as in gold printing. Bronze can be procured of various colours, and when laid on with judgment the effect is beautiful. The palest bronze is best.

PRINTING IN COLOURS.

When red and black are to be printed on the same sheet,—the same process being applicable to all other colours,—the form is made ready in the usual way, and a chalk-line is traced around the outside of the chase on the press-bed, to show the exact situation in which the form must be replaced after having been lifted. The form is then laid with its face downward on a letter-board covered with the press-blankets. The words marked in the proof to be printed red are then forced down, and Nonpareil reglets nicely fitted into the vacancies, which raise the red lines and words an equal distance from the other matter. A sheet of paper is then pasted on the form, to keep the Nonpareil underlays in their proper places. The form is again laid on the press, observing the utmost care in placing it in its original position as indicated by the marks before made on the bed.

It must then be made perfectly fast to the corner irons, as it is highly important that it remain firm and immovable during its stay on the press. The frisket (which is covered with strong paper) is then put on, the form rolled over with the red ink, and an impression made on it. The red words are then cut out with a sharp-pointed penknife, with so much nicety as not to admit the smallest soil on the paper from the other matter.

The red being finished and the form washed, the compositor unlocks it (this should be done on the imposing-stone, as the pressman can easily lay it agreeably to the marks made on the press), and draws out the red lines, filling up the space with quadrates. When this is done, the pressman cuts out the frisket for the black. An extra pair of points are used to prevent the black from falling on the red, or, as it is technically termed, *riding*. Generally, when a great number is to be printed, as many forms are used as there are colours to be printed. Another method of placing the underlays is adopted for broadsides, &c. with large letter and with but two or three lines of red. The red lines are taken out on the press, and underlays are put in, upon which the lines are placed, and the frisket is cut out as before mentioned.

The custom of printing broadsides, &c. with several colours is so common that ink-makers generally now manufacture coloured inks; consequently the printer can be supplied without the delay and labour of making. We give the following particulars, however, for the benefit of those who wish to prepare their own colours.

Varnish is the common menstruum adopted for all colours in printing. Red is the colour generally used with black. Trieste or English vermilion, with a small portion of lake, produces a beautiful red, which should be ground with a muller on a marble slab till it be perfectly smooth. If it be in the smallest degree gritty, it clogs the form, and consequently produces a thick and imperfect impression; no pains should, therefore, be spared to render it perfectly smooth; it may then be made to work as clear and free from picks as black. A cheaper red, but not so brilliant, may be prepared with orange mineral, rose pink, and red lead.

Prussian blue makes also an excellent colour, but will require a good deal of time and labour to make it perfectly smooth. It is also ground with the best varnish, but made considerably thicker, by allowing a greater portion of colour with the same quantity of varnish, than the red; it will then work clear and free from picks. As this colour dries rather rapidly, the rollers should be frequently washed.

Other colours may be made,—viz. lake and Indian red, which produce a deep red; verditure and indigo, for blues; orpiment, pink, yellow ochre, for yellows; verdigris and green verditure, for green, &c. All these colours should be ground with soft varnish, being in themselves dryers, or they will choke up the form. The consistency of the ink must be governed by the quality of the work to be executed. For a posting-bill or coarse job, the ink should be very thin, the proportion of varnish being much greater than required for fine work. Should the work be a wood-cut, or small type, the pigment should be made as thick as possible.

The best colours for printing are those of the lightest body and brightest colour.

HOW TO USE DRY COLOURS.

To produce fine qualities of coloured printing inks by mixing pure dry colours with varnish, the printer will do well to give heed to the following particulars:—

1. No more should be mixed at a time than will be required for the job in hand.

2. Coloured inks should be mixed upon a slate or marble slab, by means of the muller, and never upon an iron or other metallic table. The table, before mixing, should be thoroughly clean, and perfectly free from the slightest soil or trace of other inks.

3. For working coloured inks, the roller should not be too hard, and should possess a biting, elastic face. When change of colour is required, it should be cleaned with turpentine, and a moist sponge passed over the face, allowing a few minutes for the roller to dry before resuming its use.

For bronze printing, the roller should have a firm face, or the tenacity of the preparation may destroy it; yet it must have sufficient elasticity to deposit the preparation freely and cleanly on the type.

4. Various shades may be produced by observing the following directions:—

Bright Pink Ink.—Use carmine or crimson lake.

Deep Scarlet.—To carmine add a little deep vermilion.

Bright Red.—To pale vermilion add carmine.

Deep Lilac.—To cobalt blue add a little carmine.

Pale Lilac.—To carmine add a little cobalt blue.

Bright Pale Blue.—Cobalt.

Deep Bronze Blue.—Chinese.

Green.—To pale chrome add Chinese blue; any shade can be obtained by increasing or diminishing either colour.

Emerald Green.—Mix pale chrome with a little Chinese blue, then add the emerald until the tint is satisfactory.

Amber.—To pale chrome add a little carmine.

Deep Brown.—Burnt umber, with a little scarlet lake.

Pale Brown.—Burnt sienna; a rich shade is made by adding a little lake as above.

5. Gold Preparation. Print as with ordinary ink, then put on the bronze powder with a broad camel-hair brush; allow the impressions to remain a short time for the preparation to set, then clean off the superfluous bronze: the impressions will be much improved if passed through rollers.

HOW TO MULTIPLY COLOURS.

A printer who has on hand a stock of yellow, carmine, blue, and black inks, may produce other colours and shades by intermixing as follows:—

Yellow and carmine, mixed, will give	Vermilion.
Carmine and blue	Purple.
Blue and black	Deep blue.
Carmine, yellow, and black	Brown.
Yellow and blue	Green.
Yellow and black	Bronze green.
Yellow, blue, and black	Deep green.

Lighter shades may be obtained by adding proper proportions of white ink.

CONTRAST OF COLOURS.

It is wrongly supposed that the art of arranging colours so as to produce the best effects in printing is entirely dependent on the taste of the operator; for harmony is determined by fixed natural laws. The increasing demand for decorative or ornamental work renders it of some importance to the letter-press printer to make himself acquainted with these laws; as, without some attention to them, the most elegant designs of the type-founder, and the finest inks that can be made, may yield but an indifferent, if not a decidedly unpleasing, result.

The following remarks will be of use to persons to whom the subject is new; but for a thorough explanation of it they should refer to Chevreul on *Colours*,—a valuable work in the French language, which has been translated into English.

I. We may, in the first place, consider WHITE LIGHT as composed of three primary colours—blue, red, and yellow—duly blended; these three, in an infinite variety of proportion, serving to produce all the hues in creation. If we take any two of these primaries and mix them, we have a *secondary* colour. Thus, blue and red form *violet*, blue and yellow give *green*, red and yellow make *orange*. Each of these secondary colours harmonizes perfectly with the primary which does not enter into its composition. Violet, for instance,—itself a mixture of red and blue,—harmonizes with yellow; green, having no red in its composition, agrees well with red; orange, in the same way, forms a perfect contrast with blue. Either of these contrasts has the effect of mutually brightening the colours employed; a red and a green, &c. being more beautiful when placed side by side than when viewed singly. This is termed the HARMONY OF CONTRAST OF COLOURS; and a good example of it is seen in the scarlet geranium, or the holly; the one showing a light green leaf opposed to a bright red flower, and the other a deep green leaf with a dark red berry.

The *mixing* of colours is a very different thing from *contrasting* them; for, strange as it may seem, although one combination of the primary colours gives *white*, yet another proportion will produce *black*. While, then, red and green look beautiful side by side, it does not generally answer to print red ink on *green paper*. The reason is, that as the ink is slightly transparent, some of the green shows through it, and appears somewhat black, and thus lowers the brilliance of the red in the same degree as so much black ink would, if mixed with it. This remark will apply to orange or yellow on a blue paper, &c. The darker and fuller the body of colour used, the less it is affected in this manner.

The most perfect contrasts are those above mentioned, which are formed by the complementary colours; yet the primaries blue, red, and yellow also agree well together. But if such colours as are not in harmony are placed near each other, the effect is very damaging to their brightness. While red is made more brilliant by the proximity of green, it is dimmed and spoiled by placing it next an orange. Neither blue nor red contrasts well with violet, because the latter contains each of these colours in its composition. In any case where they must come into juxtaposition, the unpleasant effect may be lessened by adding a little of the opposite colour: so, if a violet is to contrast with red, it will be well to give it a shade of blue, making it more *purple;* if, on the other hand, it is to contrast with blue or green, it should be made *redder*.

II. Colours with Black. In all contrasts, the depth of the colour is an important element, but especially so in such as are to be affected by the presence of black. In but few instances will the latter bear the neighbourhood of a very deep colour to advantage, whilst it harmonizes with the lighter ones by contrast of tone. Yellow, from its near approach to white, should always be worked "full;" orange and green should also be full, and moderately deep in tone, to contrast with black. If a blue is employed, it should be light, or it will impoverish the black and be weakened itself. A very light blue border, with a broad margin of white between it and the body of matter enclosed, will give a clean, bright look to black ink, and whiteness to the paper. A light pink (such as carmine reduced with flake white or with clear varnish) is also good; yet perhaps the preceding is preferable. Dark and heavy borders are frequently a positive injury to printing, where the working in a light shade would have secured a good effect; for the border should always be so far secondary to the matter enclosed as not to draw off the attention too much to itself.

III. Colours on Tinted Papers and Tinted Grounds. Besides the kind of harmony already mentioned, there is another, which is produced by the contrast of light and dark shades of the *same* colour. This might be employed in letterpress more frequently than it is at present, with some advantage, as the effects it is capable of yielding are very chaste and pleasing. In a photograph or an engraving, all the effect is dependent on difference of tones of one colour; and the beauty of a wood in summer consists chiefly in the contrast displayed by a variety of shades of green only. A deep green ink on a paper of a light tone of the same colour is especially good, if a heavy letter is used; and indeed in most printing in colours, full, solid-faced letter should be preferred to outlines or shaded ones, which are difficult to work, and have at best but an inferior appearance unless the darkest tones are employed. A deep blue on a light blue ground, or against a light blue border, is also good; and without the latter accompaniment it is not unpleasant on a blue wove writing-paper. To secure the proper effect, however, the tints should be of the same *hue;* that is, if the groundwork is of a bluish green, the colour that is to be worked upon it should also be a green inclining to blue; if, on the other hand, the ground is of a yellower green, the body of ink should also be yellower; and so on. This may easily be managed by adding a small portion of ink of the colour required, until the hue is matched.

IV. Neutral Tints. In selecting borders for the more chaste description of printing, it is a pretty safe rule to avoid such as cover much surface, if they are to be worked in any strong colour or in black. When lighter tints are used, they will bear extension over a larger surface; and in this case a pale gray or neutral border will have a beneficial effect on any body with which it is contrasted, as well as on black itself, which is purified by its proximity. If the central printing is in black *only*, or in black and yellow, a *lavender* gray may be substituted for the border. And in any case in which the central matter is all in one colour, it will improve it to have a border of gray which is *slightly tinged with the complementary of such colour.* Thus, if the body be red, a very small portion of *green* may be added to the gray; and so forth.

It must be remembered that in ornamental printing absolute cleanliness is indispensable. The same roller should never be used for different colours, even after it has been washed. Instead of hanging exposed to dust and to the air, rollers should be kept in a tightly-closed box; and in this manner they will remain a long time in good order. The tins of ink should be similarly preserved, and the lids never left off except at the moment of using from them. These are small matters; but it is only by patient attention to minute details that excellence can be attained in printing.

HOW TO TREAT WOOD TYPE.

To prevent warping, all very large wood type should be set up on the edge when put away, so that both sides may be equally exposed to the air. In cleaning it, neither ley nor water should be employed under any circumstances. Turpentine, camphene, benzine, or kerosene oil may be used; but turpentine and camphene are the best. Procure a small, shallow pan: lay the form flat on a board; pour about six tablespoonfuls of turpentine into the pan; touch the face of the brush to the turpentine, and pass it quickly over the form before it evaporates. Six to eight spoonfuls of fluid will be found sufficient to clean a large form, if thus used.

WAREHOUSE DEPARTMENT.

THE WAREHOUSEMAN.

THE warehouseman should be a man sober and upright, and thoroughly competent to the business, on whom entire reliance may be placed,—one who will act upon the principle of making his employer's interest the end of all his action. The employer or foreman should frequently look to the concerns of the warehouse, and see that all the work is forwarded with despatch and accuracy.

The warehouseman should be provided with a book, termed "The Warehouse Book," with pages annexed, on the following plan, and about the size of a foolscap quarto.

Bouvier's Astronomy. (No. printed, 1000.)				
Date.	Receipt of Paper, and of whom.	No. of Copies delivered.	To whom delivered, with his signature.	For whom.
1865. Nov. 3.	90 reams of C. Magarge & Co.			
Dec. 8.	40 ditto.			
" 24.	....	100	John Troxell.	Sower, Barnes & Potts.
" 30.	....	300	R. Williams, binder.	
1866. May 4.	....	400	Edward Hughes.	"
" 5.	with waste.	230	John Warwick.	"
		1030		

When the paper is brought, the warehouseman should at once compare it with the bill of delivery, and, if right, enter the quantity immediately into the warehouse book. The number of printed copies delivered to the binder or publisher should also be entered, and his signature be taken at the time of delivery. This plan will prevent disputes with the bookseller or author relative to the receipt of paper or the delivery of sheets.

Having entered the receipt of the paper, the warehouseman should then write on each bundle, with red chalk, the title of the book it is to be used for, and remove it into a convenient part of the warehouse, or into a store-room provided for that purpose.

GIVING OUT PAPER TO WET.

A BUNDLE of paper consists of two reams, or forty quires, each quire containing twenty-four sheets. Formerly, the two outside quires were called cassie quires, as they were mostly made up of torn, stained, wrinkled, or otherwise imperfect sheets. At present, all the quires are considered good, although some outer sheets are injured by the twine used in tying up the bundles.

It is the general custom to print of every work what is termed an *even* number,—either 250, 500, 750, 1000, &c. These quantities are given out for the wetter in *tokens*,—viz.: for 250 (sheets), one token, containing 10 quires 18 sheets; for 500, two tokens, one 11 quires, and the other 10 quires and a half; for 750, three tokens, two of them 11 quires each, and the other 10 quires 6 sheets; and for 1000, four tokens, three of them 11 quires each, and the other 10 quires. If a work is printed in half-sheets, it, of course, requires only half the above quantities.

It would be difficult to form any positive and invariable rule for the quantity to be given out for short numbers, as it must depend in some degree upon the quality of the paper. The more expensive papers, on which, generally, short numbers or fine copies are printed, must be given out more sparingly than common paper, and the tympan and register sheets be supplied by a more common sort, cut to the size of the finer. For numbers up to 150, on ordinary paper, six sheets over will, generally speaking, be sufficient.

In giving out paper for what are termed *jobs*, the amount

necessary can easily be found by a simple calculation in division.

For example, a job (label or any thing else), 750 number, 32 on a sheet, will require 24 sheets, which will give an overplus of 18. Where a sheet has to be cut into many parts, allowance must be made for accidents. The overplus sheets are allowed for tympan-sheets, register-sheets, and other incidents, such as bad sheets, faults committed in rolling, pulling, bad register, &c.; in any of these casualties, the pressman doubles the sheet in the middle and lays it across the heap. In laying out the paper, the warehouseman reverses every other token, to enable the wetter to distinguish the different tokens. When this is done, he labels the heap, thus: *American Printer, May* 25, 1866,—that the pressman may know how long it has been wet, and the state it is in for working.

```
32)750(23
   64
   ---
   110
    96
   ---
    14
```

HANGING UP PAPER TO DRY.

When the paper is worked off and counted, the warehouseman carries the heap to the drying-room, where the poles are fixed for the purpose of hanging the sheets upon to dry, and lays it down on a table of convenient height, with one end of the heap toward him. He then takes the handle of the peel in one hand, and lays the top part down upon the heap, so that the upper edge may reach near the middle of the sheet; then, with the other hand, he doubles over as much of the printed paper as he thinks sufficient to hang up at one lift, which should be about twelve sheets, according to the pole-room to hang them.

In hanging up the lifts, he places them so that each lift will lap about an inch over the preceding one. It is necessary, where the end of the pole is exposed to a strong current of air from a window, to *lock* the last lift. This is done by folding a lift two or three times, so as to concentrate its weight in a small compass, and hanging this over the last lift near the window.

TAKING DOWN SHEETS WHEN DRY.

When the sheets are sufficiently dry, the warehouseman takes his peel and begins with the last lift hung up, on account of the wrapper being with that lift, and proceeds in the reverse order of hanging them up, successively taking them down, and brushing them, if dusty, till he has finished the whole.

Another way of taking the sheets down from the poles is, to lay the flat side of the peel against the edge of that lift which hangs over the other sheets, and push the peel forward, forcing them to slide, one lift over another. But by this method the dust which settles on the sheets while hanging is rubbed in.

FILLING IN AND PRESSING SHEETS.

WHEN the sheets are taken down, the warehouseman removes them to the warehouse, where they are filled in between smooth pasteboards made for the purpose. This operation is generally performed by boys, who, after a little practice, become exceedingly expert at it. We shall endeavour to be somewhat minute in our description of this operation. We will suppose the pasteboards to have sheets between them, which will be the case after they have been once used. The warehouse being provided with long tables or benches, secured to the wall, and a sufficient number of movable tables about the size of the largest paper, the warehouseman places one of the small tables endwise against the long one, forming a right angle, upon which to lay the pressed sheets as they come out of the boards; the boy then takes his stand at the right side of the table, with the dry unpressed sheets at his right hand and the pasteboards at his left, somewhat elevated, leaving sufficient space before him to fill in the sheets. He then proceeds as follows. He first moistens the thumb of his right hand and reaches across to the pasteboards at his left, drawing one off with his thumb and placing it before him; he then catches a sheet of the dry paper also with his right hand and places it as near the centre of the pasteboard as possible; then, twisting his body nimbly round to the left, he slides the pressed sheet from the pile of pasteboards to the table at his left side, and, in resuming his former position, again draws off a pasteboard with his thumb; and so on, till the gross or bundle is filled. It is then laid aside, and another bundle filled and laid across the former, taking care always to keep the bundles separated until they are put in press, when they are separated by smooth boards made of cherry or other hard wood. The bundles being all filled in, the warehouseman proceeds to fill up the standing-press, putting in one bundle at a time and placing a pressing-board between them; there should also be a stout plank introduced between the top board and the platen. In case the press should not hold quite as much as desired, more may be got in by unscrewing the press after it has been once screwed down. The press is finally screwed down as tight as possible. It should remain so for at least twelve hours, when it should be entirely emptied before the sheets are taken out of the boards. Care should be taken to keep the sides of the piles or heaps perfectly even.

COUNTING OUT AND PUTTING AWAY SHEETS.

When the sheets are taken out, the warehouseman knocks them up, and, after counting them into quires, ties them up in wrappers, marking the name of the work and signature on each bundle. Two or three sheets of each signature should be laid aside, in case the author, bookseller, or employer should want a copy of the work or a specimen of as many sheets as are finished.

STANDARD SIZES OF MACHINE-MADE PAPER.

(*Furnished by Charles Magarge & Co., Philadelphia.*)

PRINTING PAPER.

Double Imperial......inches,	32 × 44	Imperial..................inches,	22 × 32
Double Super Royal............	27 × 42	Super Royal..........................	21 × 27
Double Medium....................	24 × 38	Royal....................................	20 × 25
Royal and Half....................	25 × 30	Medium................................	19 × 24
Imperial and Half...............	32 × 33		

WRITING PAPER.

Folio.......................inches,	17 × 22	Packet Post..........inches,	11¼ × 18¼
Demy....................................	16 × 21	Commercial Post..............	11 × 17
Crown..................................	15 × 19	Letter..................................	10 × 16
Flat Cap...............................	14 × 17	Packet Note.......................	9 × 11¼
Foolscap..............................	13 × 16	Note....................................	8 × 10

A TABLE

For ascertaining the Number of Forms for a Book of any Size, and the Quantity of Paper necessary to print a thousand copies in any form, from Octavo to 36mo, half-sheetwise.

No. of Forms.	8vo.	12mo.	16mo.	18mo.	24mo.	32mo.	36mo.	Paper for 1000 Copies.	
	PAGES.	PAGES.	PAGES.	PAGES.	PAGES.	PAGES.	PAGES.	REAMS.	QRS.
1	8	12	16	18	24	32	36	1	2
2	16	24	32	36	48	64	72	2	4
3	24	36	48	54	72	96	108	3	6
4	32	48	64	72	96	128	144	4	8
5	40	60	80	90	120	160	180	5	10
6	48	72	96	108	144	192	216	6	12
7	56	84	112	126	168	224	252	7	14
8	64	96	128	144	192	256	288	8	16
9	72	108	144	162	216	288	324	9	18
10	80	120	160	180	240	320	360	11	
11	88	132	176	198	264	352	396	12	2
12	96	144	192	216	288	384	432	13	4
13	104	156	208	234	312	416	468	14	6
14	112	168	224	252	336	448	504	15	8
15	120	180	240	270	360	480	......	16	10
16	128	192	256	288	384	512	......	17	12
17	136	204	272	306	408	......	......	18	14
18	144	216	288	324	432	......	......	19	16
19	152	228	304	342	456	......	......	20	18
20	160	240	320	360	480	......	......	22	
21	168	252	336	378	504	......	......	23	2
22	176	264	352	396	......	......	......	24	4
23	184	276	368	414	......	......	......	25	6
24	192	288	384	432	......	......	......	26	8
25	200	300	400	450	......	......	......	27	10
26	208	312	416	468	......	......	......	28	12
27	216	324	432	486	......	......	......	29	14
28	224	336	448	504	......	......	......	30	16
29	232	348	464	......	......	......	......	31	18
30	240	360	480	......	......	......	......	33	
31	248	372	496	......	......	......	......	34	2
32	256	384	512	......	......	......	......	35	4
33	264	396	528	......	......	......	......	36	6
34	272	408	544	......	......	......	......	37	8
35	280	420	560	......	......	......	......	38	10
36	288	432	576	......	......	......	......	39	12
37	296	444	592	......	......	......	......	40	14
38	304	456	608	......	......	......	......	41	16
39	312	468	......	......	......	......	......	42	18
40	320	480	......	......	......	......	......	44	

EXAMPLE.—*How many reams will be required for a 12mo book containing 408 pages?* Find the number of pages (408) in the 12mo column: in the outer column on the left of the table the number of forms is seen, and in the outer column on the right the quantity of paper required is given.

ORTHOGRAPHICAL.

A THOROUGH reformation of the orthography of the English language, desirable as it is, can scarcely be hoped for in this century; though doubtless the time will come when an international convention will settle authoritatively the spelling of every word, as acceptably as has been done by the Academies of France and Spain in regard to the orthography of the languages of those countries. We therefore note only some discrepancies in English spelling, and indicate what appears to be the preferable method.

A or AN *before a Vowel or silent* H.

In regard to the use of the indefinite article, Walker's Dictionary very judiciously says,—

"This indefinite, and, as it may be called, the *euphonic* article, is said by all our grammarians to be used before a vowel or *h* mute; but no notice is taken of using *a* instead of it before what is called a vowel, as, *a useful book*, *a useful ceremony*, *a usurer*, &c.; nor is any mention made of its constant usage before *h* when it is not mute, if the accent of the word be on the second syllable, as, *an heroic action*, *an historical account*, &c. This want of accuracy arises from a want of analyzing the vowels, and not attending sufficiently to the influence of accent on pronunciation. A proper investigation of the power of the vowels would have informed our grammarians that the letter *u*, when long, is not so properly a vowel as a semi-consonant, and perfectly equivalent to commencing *y*, and that a feeling of this has insensibly influenced the best speakers to prefix *a* to it in their conversation, while a confused idea of the general rule, arising from an ignorance

24

of the nature of the letters, has generally induced them to prefix *an* to it in writing. The same observations are applicable to the *h*. The ear alone tells us that, before *heroic*, *historical*, &c., the *an* ought invariably to be used; but, by not discovering that it is the absence of accent on the *h* that makes *an* admissible in these words, we are apt to prefix *an* to words where the *h* is sounded, as, *an horse*, *an house*, &c., and thus set our spoken and written language at variance. The article *a* must be used before all words beginning with a consonant, and before the vowel *u* when long; and the article *an* must be used before all words beginning with a vowel, except long *u*; before words beginning with *h* mute, as, *an hour*, *an heir*, &c.; or before words where the *h* is not mute, if the accent be on the second syllable, as, *an heroic action*, *an historical account*, &c." The few words in our language in which the *h* is mute are *heir*, *herb*, *honest*, *honour*, *hospital*, *hostler*, *hour*, *humble*, *humour*, and their derivatives.

ABLE *and* IBLE.

All English words, without regard to the source from which they have been derived, and those which come from Latin words ending in *abilis* or French ones in *able*, take the termination *able* in English, as, *procurable*, *amendable*, *desirable*, *allowable*, *voidable*, *available*, *fordable*, *incontestable*, &c.; but in words from Latin and French words terminating in *ibilis* or *ible*, then the ending will be *ible* in English. For instance: *accessible*, *sensible*, *defensible*, *convertible*, &c.

In words ending in *ce* or *ge*, the final *e* is preserved before the termination *able*, for the purpose of indicating the soft sound of the consonant, as in *marriageable*, *chargeable*, *traceable*, *serviceable*, &c.; but before the ending *ible* the final *e* of the primitive disappears, and there is no *e* before the termination. Examples: *deducible*, *reducible*, *frangible*, &c.

The following list of words in *ible* is here added; all others end in *able*.

accessible	coercible	conceptible	convincible
admissible	collectible	conclusible	corrigible
adustible	comminuible	congestible	corrosible
appetible	compatible	contemptible	corruptible
apprehensible	competible	contractible	credible
audible	comprehensible	controvertible	deceptible
cessible	compressible	convertible	decerptible

decoctible	eligible	immiscible*	referrible
deducible	eludible	impassible†	reflexible
defeasible	enforcible	intelligible	refrangible
defectible	evincible	irascible	regible
defensible	expansible	legible	remissible
depectible	expressible	miscible	reprehensible
deprehensible	extendible	partible	resistible
descendible	extensible	passible‡	responsible
destructible	fallible	perceptible	reversible
digestible	feasible	permiscible	revertible
discernible	fencible	permissible	risible
discerptible	flexible	persuasible	seducible
dispraisible	forcible	pervertible	sensible
dissolvible	frangible	plausible	solvible
distensible	fusible	possible	tangible
divisible	horrible	producible	terrible
docible	ignoscible	quadrible	transmissible
edible	illegible	reducible	visible
effectible	immarcessible		

IM *or* IN *and* EM *or* EN.

The prefix *in* is from the Latin, and that of *en* from the French and Greek. *In* generally signifies *situation*, and *en* mostly expresses *action*. Hence, perhaps, in strictness, *inclose* will signify "to close in," and *enclose*, "to make close." So, to *inquire* will be "to seek *in*, or to search in," and *enquire*, to "make search." *Immigrate*, "to pass into;" *emigrate*, "to go out of." But this distinction is not attended to by writers, and is, indeed, too refined for general practice.

Before the letters *b* and *p*, *en* becomes *em*, as in *embattle*, *empower;* and *in* before some letters becomes *ig*, *il*, *im*, or *ir*, as in *ignoble*, *illegal*, *improper*, *irresolute*.

We give a list of those generally spelt with *im* or *in;* leaving it to be inferred that the rest are more usual with *em* or *en*.

imbarn	imbue	immigrate	impale	impel
imbibe	imburse	immingle	imparadise	impen
imboil	immanacle	immit	impassioned	imperil
imbound	immense	immix	impawn	impinge
imbrue	immerge	immure	impeach	implant
imbrute	immerse	impact	impearl	implead

---

* For other words beginning with *im*, *in*, *ir*, or *un* negative, look for the simple word.

† Incapable of suffering. ‡ Capable of suffering.

import	indict	ingest	insert	intrench
impose	indite	inhabit	inset	intrude
impound	indoctrinate	inhale	inshell	intrust
impregnate	indrench	inhere	inship	inumbrate
impress	induce	inhold	insinew	inure
imprint	induct	inhume	insphere	inurn
imprison	ineye	initiate	inspire	invade
inarch	infer	inject	inspirit	inveigh
incase	infest	inlapidate	install	invert
inclasp	infix	inlay	instate	invest
inclip	inflame	inlet	insteep	invigorate
incloud	inflate	inoculate	instil	invite
include	inflect	inosculate	instop	invocate
incrassate	inflict	inquire	insure	invoice
increase	infringe	inrail	inter	invoke
incur	infuscate	inscribe	intertwine	inwall
indart	infuse	insculp	intort	inweave
indent	ingrain	inseam	intreasure	

IN *and* UN.

In, as a prefix, also marks *negation:* it is probable that it came from the Romans. *Un*, as a prefix, is synonymous with *in:* it is of Saxon origin, and generally joined to words from a northern source; while *in* is oftener applied to those of Latin derivation.

ISE *and* IZE.

The variation in the terminations *ise* and *ize* is due to the different derivations of words,—*ize* characterizing words from the Greek and Latin, and *ise* from the French. The rule, however, is not inflexible. The following words are commonly spelled with the *s*.

advertise	comprise	divertise	merchandise
advise	compromise	emprise	misprise (mistake)
affranchise	criticise	enfranchise	premise
aggrandise	demise	enterprise	recognise.
amortise	despise	exercise	reprise (take again)
catechise	devise	exorcise	supervise
chastise	disfranchise	galliardise	surmise
circumcise	disguise	manumise	surprise

OR *and* OUR.

The ending *our* was in general use until the appearance of Webster's Dictionary, in which the *u* was dropped in words terminating with *our*. This innovation has steadily gained ground. We are not disposed to approve of partial tinkerings with English orthography; and, until a general convention of British and American scholars settle the method of spelling English words, we shall adhere to the established usage. We append a list of words terminating in *our*.

arbour	colour	fervour	odour	splendour
ardour	contour	flavour	parlour	succour
armour	demeanour	harbour	rancour	tambour
behaviour	dishonour	honour	rigour	tumour
candour	dolour	humour	rumour	valour
clamour	endeavour	labour	savour	vapour
clangour	favour	neighbour	saviour	vigour

The *u* is dropped when the termination *ous* is added to any of these words; as, *clamorous*, *dolorous*, *humorous*, *laborious*, *odorous*, *rancorous*, *rigorous*, *valorous*, *vigorous*. And also in derivative words; such as *armory*, *honorary*, &c.

SION *and* TION.

Primitive words which end in *d*, *de*, *ge*, *mit*, *rt*, *se*, or *ss*, take *sion* in their derivatives; but all other words have *tion*.

EXAMPLES.

absin*d*, abscission	rever*t*, reversion
condescen*d*, condescension	conver*t*, conversion
eva*de*, evasion	confu*se*, confusion
intru*de*, intrusion	revi*se*, revision
absterg*e*, abstersion	impre*ss*, impression
emer*ge*, emersion	confe*ss*, confession
ad*mit*, admission	admi*x*, admixtion
re*mit*, remission	promo*te*, promotion

IRREGULARS.

adhesion	divulsion	recension	attention
cohesion	evulsion	recursion	causation
compulsion	exesion	revulsion	distention
declension	expulsion	scansion	distortion
decursion	impulsion	tension	coercion
depulsion	incursion	transcursion	suspicion
dissension	propulsion	version	crucifixion

FARTHER *and* FURTHER.

Farther is nowadays only employed when speaking of *distance;* in all other acceptations of the word, *further* is generally adopted.

PEAS *and* PEASE.

There are scarcely any words in which a mistake is more frequently made than in *peas* and *pease*. Yet the distinction between them is simple and well defined. *Peas* is the plural of *pea*, and, consequently, only follows *numeral* adjectives; as, "*ten* peas," "a *hundred* peas," "a *few* peas," "*many* peas;" but *pease* is used when speaking of the legumen in the aggregate, or generally. Thus, we correctly say, "*Pease* are dear this year," "*Pease* were plentifully supplied to the horses," &c.

Pease is also employed adjectively; as, "*pease*-pudding," "*pease*-soup," or "*pea*-soup," &c.

The Omission of s *in the Possessive Case.*

It is not uncommon with some persons to omit the *s* after the apostrophe in the possessive case of nouns, if the name itself ends in *s;* as, "*James*' book," "*Barnes*' Notes." But this is incorrect; for if we ask, Whose book? we should directly answer, *James's*. The only case when the *s* can be judiciously omitted, and this solely to avoid the too hissing sound of so many *s*'s in succession, is when the first word ends with the sound of *s* in its last two syllables, and the next word begins with *s;* as in *Misses*' spectacles, *righteousness*' sake, *conscience*' sake.

Formation of the Plurals of Words compounded of a Noun and an Adjective.

Adjectives have no plural number. Therefore, in a word compounded of a noun and an adjective, the *s* denoting the plural number is attached to the end of the noun, as follows:—

Governor-general................................	Governors-general.
Attorney-general................................	Attorneys-general.
Court-martial....................................	Courts-martial.

But where the adjective is taken substantively, the mark of the plural will properly follow it. For example: *Brigadier-generals*, *major-generals*, *lieutenant-generals*.

Words compounded of a noun and the adjective *full* form their plurals thus: *spoonfuls, cupfuls, bucketfuls, handfuls, mouthfuls.*

Pointing of Numbers, Weights, Measures, &c.

No comma should be placed between the constituent parts of the same number, however long it may be. Thus, we say, "One million one hundred thousand five hundred and twenty-one," without any interpunction. The reason is, that there is no more than *one* numerical aggregate intended, or but *one* complex notion; and, consequently, no separation of parts or members can take place. The same reasoning holds good as respects *values, weights, &c.* For instance, when we say, "Six dollars and ten cents," we merely mean that aggregate amount, but not necessarily any one of the coins indicated. If we did so intend, then two commas should be introduced,—one after "dollars," and the other after "cents." In like manner we should act with such sentences as, "Five tons three hundred-weight two quarters and fifteen pounds;" or, "Ten acres four roods and twenty-seven perches;" and for the same reason: no division of parts is intended, but merely one aggregate amount.

When figures are used to express amounts, a comma should not be inserted to cut off the tens unless the sum requires five figures: *e.g.* $10,600, 20,000 men, &c. In column matter this rule will not apply.

Derivation of English Words.

Of course the Saxon forms the basis of our language in its essential parts, and is the source whence we derive the greater part of our ordinary and most emphatic words. Nevertheless, various other languages have been put under contribution, especially the French, Latin, and Greek. This will be evident from the following statement of derivations, which will show the unlearned reader how important it is to him that he should acquire some knowledge of those languages, if he desires to attain to a thorough proficiency in his business as an educated printer.

I. From the Greek are derived—

1. Words ending in *gram, graph,* and *graphy;* as, *telegram,*

telegraph, *geography*, &c.; from the word γράφω (*grapho*), I write, and some other Greek word.

2. Those in *gon;* from γωνία (*gonia*), an angle; as, *octagon.*

3. All words in *logue* or *logy;* as, *epilogue*, *astrology;* from λόγος (*logos*), a discourse.

4. *Ic*, *ick*, *ics* are also Greek terminations, generally of adjectives.

5. Words in *meter* are all of Greek origin, coming from the verb μετρῶ (*metro*), I measure, in combination with some other word.

6. Most words into which the terminations *agogue*, *asis*, *esis*, or *ysis* enter are also of Greek origin; such as *demagogue*, *emphasis*, *parenthesis*, *analysis*, &c.

II. But the main source whence we have derived words, with the exception of the Saxon, is the Latin, as will appear from an inspection of the following list:—

1. Words ending in *ance*, *ancy*, or *ant*, and *ence*, *ency*, or *ent*, come from Latin words ending respectively in *ans*, *antia*, or *ens*, *entia;* as, *abundance*, from *abundantia;* *infancy*, from *infantia;* *abundant*, from *abundans;* *absence*, from *absentia;* *excellency*, from *excellentia;* and *excellent*, from *excellens.*

2. Words in *al* have their Latin representatives in *alis;* as, *corporal*, from *corporalis.*

3. Verbs in *ate* mostly come from Latin verbs of the first conjugation; as, *moderate*, from *modero.*

4. Words in *ator* are generally the same in both languages; as, *orator*, *senator*, *moderator.*

5. The termination *id* comes mostly from Latin words ending in *idus;* as, *acid*, from *acidus;* but sometimes words of this ending are of Greek origin; as, *oxide* (more correctly, *oxyd*), from ὀξὺς (*oxys*); and, indeed, most scientific words of this ending; as, *carotid*, from καρώτιδες, &c.; *rhomboid*, from ῥομβοειδής.

6. *Il* or *ile* is likewise from the Latin termination of adjectives in *ilis;* as, *docile*, from *docilis;* civil, from *civilis.*

7. The Latin termination *osus* has its English representative in *ious* or *ous;* as, *copious*, from *copiosus;* *numerous*, from *numerosus.* But sometimes the English ending *ous* comes from a Latin word in *ax;* as, *capacious*, from *capax.*

8. The Latin ending *io* has its English corresponding word in *ion;* as, *nation*, from *natio;* *oration*, from *oratio.*

9. The endings *ne*, *re*, and *te* after a vowel are also for the greater part of Latin origin; as, *fortune*, from *fortuna; aquiline*, from *aquilinus; culture*, from *cultura; pure*, from *purus, complete*, from *completus*, &c.

10. Words in *ty* come from Latin words in *tas;* as, *equality*, from *æqualitas; bounty*, from *bonitas; rarity*, from *raritas*, &c.

11. The termination *ude* is also of Latin origin, coming from words in *udo;* as, *fortitude*, from *fortitudo; elude*, from *eludo.*

12. So also is *uous*, by inserting the letter *o;* as, *ambiguous* from *ambiguus; continuous*, from *continuus*, &c.

III. From the French have come—

1. Most of our words in *age;* as, *page*, *rage*, *usage.*

2. All those in *eau;* as, *beau*, *flambeau*, &c.

3. The French *esse* is represented by the English *ess;* as, *princess*, from *princesse.*

4. Words in *que* mostly come to us from the French directly; some from the Latin directly or indirectly; as, *antique* (L. *antiquus*, F. *antique*), *oblique*, *opaque.*

5. Words ending in *ment* are nearly the same in both languages; as, *commencement*, *advancement* (F. *avancement*), &c.

We subjoin some rules for spelling, adapted from Laidlaw's *American Pronouncing Dictionary.**

RULE I.

Words ending in silent *e* after *u* or a consonant generally drop the *e* on taking an additional termination beginning with a vowel; as, sale, *salable;* plague, *plaguy;* sue, *suing;* eye, *eying.*

EXCEPTION I.—Words ending in *ce* and *ge* retain *e* before *able* and *ous;* as, service, *serviceable;* trace, *traceable;* courage, *courageous;* advantage, *advantageous.*

EXCEPTION II.—Compounds and prefixes retain *e;* as, *firearms*, *foreordain*, *poleaxe*, *vice-admiral*, *fire-engine.*

---

* Published by E. C. Markley & Son, Philadelphia. An excellent book.

Remark.—From singe, springe, swinge, tinge, we write *singeing*, *springeing*, *swingeing*, *tingeing*, to distinguish from *singing*, *springing*, *swinging*, and *tinging*. *Dyeing*, from dye, retains *e*, to distinguish it from *dying*, the present participle of die. Mile retains *e* in *mileage*. Derivatives from proper names of persons retain *e*; as, *daguerreotype*, *morseograph*.

RULE II.

Words ending in silent *e* generally retain the *e* on taking an additional termination beginning with a consonant; as, bereave, *bereavement*; issue, *issueless*.

Remark.—*Awful*, *awfully*, *awfulness*, *argument*, *argumentation*, *argumentative*, *woful*, *wofully*, *wofulness*, *duly*, *truly*, and *wholly*, are undisputed exceptions; and *abridgment*, *acknowledgment*, *judgment*, *misjudgment*, *prejudgment*, *lodgment*, *wobegone*, and *rhymster*, are disputed exceptions. Some write *abridgement*, *acknowledgement*, *judgement*, *misjudgement*, *prejudgement*, *lodgement*, *woebegone*, and *rhymester*.

RULE III.

Words ending in *ie* change them into *y* before *ing*; as, lie, *lying*. The following words conform to this rule:—

lie	lie	die	untie
belie	overlie	hie	vie
outlie	underlie	tie	outvie

RULE IV.

Words ending in *y* preceded by a consonant generally change *y* into *i* on taking an additional syllable; as, mercy, *merciful*, *merciless*; defy, *defied*, *defies*, *defieth*, *defiant*; busy, *busier*, *busiest*, *business*; ply, *pliers*; porphyry, *porphyritic*.

Exception I.—*Y* after a consonant is not changed into *i* before *ing* or *ish*; as, *dry*, *drying*, *dryish*.

Exception II.—Compounds usually retain *y*; as, *mercy-seat*, *county-town*, *dairymaid*, *skylight*.

Remark.—*Dryer*, *dryest*, *dryly*, *dryness*, *shyer*, *shyest*, *shyly*, *shyness*, are undisputed exceptions to the rule; and *slyer*, *slyest*, *slyly*, *slyness*, are disputed exceptions.

RULE V.

Words ending in *y* preceded by a vowel retain the *y*; as, gay, *gayly*, *gayness*, *gayety*; pray, *prayer*, *praying*, *prayed*, *prays*.

Remark.—From day, lay, pay, say, stay, are formed *daily*, *laid*, *paid*, *said*, *saith*, *staid*. The regular words *dayly*, *layed*, *payed*, *sayeth*, and *stayed*, are sometimes used.

RULE VI.

Monosyllables and words having the primary accent on the last syllable, when they end with a single consonant preceded by a single vowel, double their final consonant before an additional syllable that begins with a vowel; as, wet, *wetter, wettest, wetting, wetted;* drum, *drumming, drummed;* dispel, *dispelling, dispelled.*

EXCEPTION.—A final *x*, or the *s* in gas, should not be doubled; as, fix, *fixes, fixed, fixing; annex, annexing; gases, gasefy.*

Remark I.—*U* after *q* is never reckoned a part of a diphthong or triphthong; so that from quit are formed *quitting, quitted;* and from quag, *quaggy.*

Remark II.—This rule applies only to derivatives which retain the accent of their primitives, and not to such as *in'ferable, in'ference, pref'erable, pref'erence, ref'erable,* and *ref'erence,* from infer, prefer, and refer. To the forms *infer'rible, refer'rible,* which are sometimes met with, the general rule applies. *Transfer'able,* from transfer, is an exception to the general rule; the regular form *transfer'rible* is not often used. Although *parallel'ogram,* from par'allel, and *modal'ity,* from mo'dal, remove the primary accent to the point of duplication, they do not double the final *l*. See Remark II. under Rule VII.

RULE VII.

A final consonant is not doubled when it is preceded by a diphthong, when the primary accent is either not on, or not retained upon, the last syllable, or when the additional syllable begins with a consonant; as, beat, *beating, beaten;* dif'fer, *dif'fering, dif'fered, dif'ference, dif'ferent;* prefer', *pref'erence;* refer', *ref'erence;* fit, *fitful, fitly, fitness; ben'efit, ben'efited, ben'efiting.*

EXCEPTION I.—Compounds that remove the primary accent from the point of duplication retain the double letter; as, *broad'-brimmed, heel'tapping.*

Remark I.—When *ly* is affixed to words ending in *l*, the l is not considered doubled; as in *cool-ly, real-ly, gravel-ly, royal-ly.*

Remark II.—*Nutmegged, kidnapping, kidnapped, kidnapper, zigzagging, zigzagged, excellence,* and some others, are undisputed exceptions to the rule. There are nearly one hundred words, from which more than four hundred derivatives are formed, that are usually made exceptions to this rule. Webster is distinguished for making nearly all the derivatives conform to the rule. Webster and Smart accent the verb *curv'et* on the first syllable, with which accentuation *curveting* and *curveted* are correct spellings; other orthoepists accent it upon the last syllable, then *curvet'ting* and *curvet'ted* are correct.

RULE VIII.

Words ending in *c* accept of *k* before a termination beginning with *e*, *i*, or *y;* as, frolic, *frolicked*, *frolicking;* colic, *colicky.*

ILLUSTRATIONS.

colic	mimic	rollic	talc
colicky	mimicking	rollicking	talcky
	mimicked	rollicked	
frolic	mimicker		zinc
frolicking		traffic	zinckiferous
frolicked	physic	trafficking	(zinciferous)
	physicking	trafficked	zincky
havoc	physicked	trafficker	
havocking			
havocked			

RULE IX.

Words ending in a double letter preserve it double after a prefix or before a termination beginning with a different letter; as, *op-press*, *mis-spell*, *in-thrall*, *oversee; see-ing*, *oppress-ive*, *stiff-ness*, *woo-ed*, *still-ness*, *assess-ment.*

Remark I.—Annul, *until*, *twibil*, and the conservative *fulfil*, or the Websterian *fulfill*, are the only exceptions to the first part of this rule extensively recognised by present usage. The conservative *distil* and *instil* are at variance; but the Websterian *distill* and *instill*, and also *twibill*, as written by Reid, are in harmony with the rule.

Remark II.—Pontific, and all other derivatives of pontiff, are exceptions to the latter part of this rule, unless an *f* is discarded in the primitive word, as Webster suggests and the derivation warrants. The derivatives of dull, full, skill, and will, are disputed exceptions: if spelled as Webster writes them, *dullness*, *fullness*, *skillful*, *willful*, they conform to the rule.

RULE X.

The plural is usually formed from the singular by adding *s;* as, brave, *braves;* night, *nights;* hymn, *hymns.*

RULE XI.

Nouns ending in *o* preceded by a vowel accept of *s* in the plural; as, cameo, *cameos;* studio, *studios.*

ILLUSTRATIONS.

agios	curculios	koodoos	punctilios
bagnios	embryos	nuncios	ratios
bamboos	folios	olios	seraglios
braggadocios	imbroglios	oratorios	solfeggios
cameos	intaglios	pistachios	studios
cuckoos	internuncios	port-folios	trios

RULE XII.

Nouns ending in *y* preceded by a vowel accept of *s* in the plural; as, money, *moneys;* attorney, *attorneys.*

RULE XIII.

Nouns ending in *o* preceded by a consonant usually accept of *es* in the plural; as, echo, *echoes;* embargo, *embargoes.*

Remark.—There are more than fifty words that conform to this rule, and about thirty that accept of *s* only.

ILLUSTRATIONS.

archipelagoes	frescoes	negroes	rotundoes
armadilloes	grottoes	palmettoes	stilettoes
bilboes	gustoes	passadoes	supercargoes
bravadoes	heroes	peccadilloes	testudoes
bravoes	innuendoes	potatoes	tomatoes
buffaloes	juntoes	prunelloes	tornadoes
buffoes	lazarettoes	punchinelloes	torpedoes
calicoes	lingoes	punctoes	umboes
cargoes	lumbagoes	ranchoes	vetoes
desperadoes	mangoes	recitativoes	violoncelloes
echoes	manifestoes	relievoes	viragoes
embargoes	mottoes	renegadoes	volcanoes
farragoes	mulattoes	ritornelloes	zeroes

EXCEPTIONS.

albinos	inamoratos	octavos	rondos
cantos	lassos	pianos	salvos
centos	limbos	porticos	set-tos
dominos	major-domos	provisos	siroccos
duodecimos	mementos	quartos	solos
halos	merinos	rancheros	torsos
hidalgos	mosquitos	ridottos	tyros

It would be well if all words ending in *o* were made to conform to Rules XI. and XIII.

RULE XIV.

Nouns ending in *ss*, *z*, *x*, *ch* soft, and *sh*, accept of *es* in the plural; as, dress, *dresses;* buzz, *buzzes;* box, *boxes;* peach, *peaches;* dish, *dishes.*

RULE XV.

Nouns ending in *y* after a consonant change *y* into *ies* in the plural; as, city, *cities;* daisy, *daisies.*

RULE XVI.

Compound nouns whose parts are connected by a hyphen accept of the sign of the plural after that part which essentially constitutes the noun; as, knight-errant, *knights-errant;* son-in-law, *sons-in-law;* man-of-war, *men-of-war;* step-child, *step-children;* ember-day, *ember-days;* man-singer, *men-singers.*

ILLUSTRATIONS.

aides-de-camp	courts-martial	knights-errant	prices-current
beaus-ideal or	cousins-german	mesdames	sergeants-at-arms
beaux-ideal	daughters-in-law	men-of-war	sisters-in-law
cartes-blanche	fathers-in-law	messieurs	sons-in-law
charges-d'affaires	gendarmes or	mothers-in-law	step-children
chevaux-de-frise	gens d'armes	poets-laureate	step-fathers
coups-de-main	jets d'eau	porte-monnaies	valets-de-chambre

Remark I.—If no hyphen is used, the sign of the plural is always placed at the end; as, spoonful, *spoonfuls.*

Remark II.—The sign of the possessive case is always placed at the end of compound nouns; as, *son-in-law's* house.

RULE XVII.

The compounds of man form their plural in the same manner as the simple word; as, fisherman, *fishermen;* man-of-war, *men-of-war.*

EXCEPTIONS.—The only exceptions to this rule are *dragoman*, *Mussulman*, *Ottoman*, *talisman*, *Turcoman*, *German*, *Normans*, and *landamman*, which accept of *s*.

RULE XVIII.

Of the terminations *eive* and *ieve*, and of the derivatives of each, the former are found after *c*, and the latter after other letters; as, *conceive, conceit, receive, receipt; relieve, relief, relieving, thieve, thievish.*

ILLUSTRATIONS.

achieve	deceitful	misbelief	reprieve
aggrieve	deceive	misconceive	retrieve
bas-relief	disbelief	perceive	sieve
belief	disbelieve	preconceive	thief
believe	grief	receipt	thieve
conceit	grieve	receive	unbelief
conceivable	inconceivable	relief	unbeliever
conceive	lief	relieve	undeceive
deceit	lieve	relievo	

Plurals of Nouns which change F *or* FE *into* VES.

beeves	knives	selves	tipstaves
calves	leaves	sheaves	wharves
elves	lives	shelves	wives
halves	loaves	thieves	wolves

All other nouns ending in *ff* conform to Rule X. Wharfs prevails in Great Britain, wharves in America.

Plurals of Nouns ending in F *or* FE *which accept of* S *only in the Plural.*

briefs	caliphs	surfs	roofs
chiefs	caufs	fifes	proofs
fiefs	clefs	strifes	reproofs
griefs	coifs	safes	disproofs
mischiefs	delfs	scarfs	waterproofs
kerchiefs	dwarfs	waifs	beliefs
neckerchiefs	turfs	woofs	reliefs
handkerchiefs	kerfs	hoofs	gulfs

Plurals of Nouns ending in EAU, IEU, *and* OU.

beaux	flambeaux	portmanteaus	bijoux
bureaux	plateaux	purlieus	morceaux
chapeaux	rondeaux	adieux	rouleaux
chateaux	jets d'eau	batteaux	tableaux

A number of these nouns admits of two forms in the plural.

THE LAW OF COPYRIGHT.

Subjects of Copyright.—The intellectual productions to which the law extends protection are of three classes. *First*, writings or drawings capable of being multiplied by the arts of printing or engraving. *Second*, designs of form or configuration capable of being reproduced upon the surface or in the shape of bodies. *Third*, inventions in what are called the useful arts. To the first class belong books, manuscripts, maps, charts, music, prints, cuts, and engravings; to the second class belong statuary, bas-reliefs, designs for ornamenting any surface, and configurations of bodies; the third class comprehends machinery, tools, manufactures, compositions of matter, and processes or methods in the arts. According to the practice of legislation in England and America, the term *copyright* is confined to the exclusive right secured to the author or proprietor of a writing or drawing, which may be multiplied by the arts of printing in any of its branches. Property in the other classes of intellectual objects is usually secured by letters-patent, and the interest is called a *patent-right*. But the distinction is arbitrary and conventional.

The Persons entitled to secure a Copyright.—Any person or persons being a citizen or citizens of the United States, or resident therein, and being the author or authors of any book, map, chart, or musical composition, or who shall invent, design, etch, engrave, work, or cause to be engraved, etched, or worked from his own design, any print or engraving, and the executors, administrators, or legal assigns of any such person or persons, may secure a copyright therein by complying with the requisitions of the statute. The term "citizen" comprehends both native-born and naturalized citizens. A foreigner may take a copyright, if resident; but whether such residence must be permanent, or may be temporary, has not been judicially determined.

THE TERM for which a copyright may be obtained is the period of twenty-eight years from the time of recording the title; and at the expiration of that period the author, engraver, or designer, if living, and his widow and children, if he be dead, may re-enter for fourteen years additional or renewed term.

THE FORMALITIES requisite to the securing of the original term are—1. The deposit of a printed copy of the title of the book, map, chart, musical composition, print, cut, or engraving, in the clerk's office of the United States district court for the district where the author or proprietor resides. 2. The recording of that title by the clerk. 3. The deposit of a copy of the book, &c. with the same clerk within three months of the time of publication. 4. The printing of a notice that a copyright has been secured on the title-page of every copy, or the page immediately following, if it be a book, or on the face, if it be a map, chart, musical composition, print, cut, or engraving, or on the title or frontispiece, if it be a volume of maps, charts, music, or engravings, in the following words:—

"Entered according to act of Congress, in the year [1866], by [L. Johnson & Co.], in the clerk's office of the District Court of [the Eastern District of Pennsylvania]."

To obtain the renewed or additional term, these requisites must be repeated within six months before the expiration of the first term; and a notice must also be printed within two months of the date of the entry for renewal, for the space of four weeks, in one or more newspapers printed in the United States, which notice must consist of a copy of the record of the renewal.

Prior to the act of Congress "providing for keeping and distributing all public documents," approved February 5, 1859, the law provided that one copy of each book or other production should be sent to the librarians of the Smithsonian Institute, and one to the librarian of the Congress Library. This provision is now repealed; and while in existence it was questionable whether a compliance with its conditions was essential to a valid copyright.*

* Bouvier's *Law Dictionary*, published by George W. Childs, Philadelphia.

TECHNICAL TERMS OF THE CRAFT.

Alley.—The space between two stands.

Ascending letters.—Letters that ascend into the upper shoulder: as, b, d, l, &c., and all the capitals.

Author's proof.—The clean proof sent to an author after the compositors' errors have been corrected.

Bank.—A table about four feet high, to lay sheets on at press.

Bastard title.—A short title preceding the general title of a work.

Bastard type.—Type with a face larger or smaller than its appropriate body: as, Nonpareil on Minion body, or Minion on Nonpareil body.

Batter.—Types accidentally injured in a form.

Beard of a letter.—The outer angles supporting the face of a type and extending to the shoulder.

Bearer.—A strip of reglet to bear off the impression from a blank page. A long piece of furniture, type-high, used in working jobs. A solid-faced type interspersed over the blank parts of a page, in composing for stereotyping, to resist the force of the knife when the plates are shaved.

Bearer-lines.—The top line and bottom line in a page prepared for stereotyping.

Bed.—The flat part of the press on which the form is laid.

Bevels.—Slugs cast nearly type-high, with a bevelled edge, used by stereotypers to form the flange on the side of the plates.

Bite.—An irregular white spot on the edge or corner of a printed page, caused by the frisket not being sufficiently cut out.

Blanket.—A woollen cloth used in the tympan.

Blank-line.—A line of quadrates.

Blocks.—The mahogany frames on which stereotype plates are affixed for printing.

Bodkin.—A delicate awl-like tool used for correcting errors in type.

Body.—The shank of the letter.

Botch.—A bungling, incompetent workman.

Bottle-arsed.—Type wider at the bottom than at the top.

Boxes.—The compartments of a case in which the types are placed.

Brayer.—A wooden or glass rubber, flat at the bottom, used to bray or spread out ink on the ink-block.

Break-line.—A short line.

Broadside.—A form of one page, printed on one side of a whole sheet of paper.

Broken matter.—Pages of type disrupted and somewhat intermingled.

Bundle.—Two reams of paper.

Bur.—Rough edge of a type which the founder neglected to take off in dressing.

Caret.—A character [∧] used to denote the place where an omission in the proof should be inserted.

Case.—The receptacle for type, divided into numerous compartments.

Cassie paper.—Formerly, the two outside quires of a ream, consisting of defective sheets.

Casting off.—Estimating how many pages a certain quantity of copy will make in type.

Ceriphs.—The fine lines and cross-strokes at the end of a letter.

Chapel.—A printing-office.

Chase.—A rectangular iron frame in which pages of type are imposed.

Clean proof.—A proof containing few faults.

Clearing away.—Properly disposing of materials after a work has been completed.

Close matter.—Solid matter with few break-lines.

Companionship.—All the hands on a work.

Composing.—Setting type.

Composing-rule.—A steel or brass rule, with a beak at one end, used in type-setting.

Composing-stick.—An instrument in which types are arranged in words and lines.

Correct.—A compositor is said to correct when he amends the faults marked in a proof.

Corrections.—The alterations or errors marked in a proof.

Cut-in letter.—A type of large size adjusted at the beginning of a line at the commencement of chapters.

Cut-in note.—A note justified into the side of a page.

Dead horse.—Matter charged before it is set.

Dele, ꝺ.—A proof-reader's mark, signifying to take out.

Devil.—The errand-boy of a printing-office.

Dished.—A defect in electrotyped plates, the centre of a letter being lower than its edges.

Distributing.—Returning type to their various boxes after having been printed from. Spreading ink evenly over the surface of a roller.

Double.—Among compositors, a repetition of words; among pressmen, a sheet that is twice pulled and mackled.

Dressing a chase or form.—Fitting the pages and chase with furniture and quoins.

Drive out.—To space widely.

Duck's-bill.—A tongue cut in a piece of stout paper and pasted on the tympan at the bottom of the tympan-sheet, to support the paper when laid on the tympan.

Duodecimo, or *12mo.*—Twelve pages to a form.

Em.—The square of the body of a type.

En.—Half the dimensions of the preceding.

Even page.—The 2d, 4th, 6th, or any even-numbered page of a book.

Fat.—Poetry and leaded matter.

Fat face, or *Fat letter.*—Broad-stemmed letter.

First form.—The form first printed, which generally contains the first page of a sheet.

Fly.—The person or apparatus that takes off the sheets from the press.

Folio.—Two pages to a form.

Foot-sticks.—Sloping pieces of furniture placed at the bottom of pages, between which and the chase the quoins are driven to fasten the pages.

Form.—The pages when imposed in a chase.

Foul proof.—A proof with many faults marked in it.

Fount.—An assortment of type in definite proportions.

Friar.—A light patch in a printed sheet, caused by defective rolling.

Frisket.—An iron frame fastened by a hinge to the upper part of the tympan, to hold the sheet of paper fast as it goes in and comes from the press.

Fudge.—To contrive without proper materials.

Full press.—When two men work at the press with hand-rollers.

Furniture.—Strips of wood or metal placed around and between pages when imposed.

Galley.—A wooden or brass flat oblong tray, with side and head ledges, for holding type when composed.

Galley-slaves.—An ancient term of derision applied by pressmen to compositors.

Gauge.—A strip of reglet with a notch in it, passed with the make-up, to denote the length of the pages.

Get in.—To set close.

Good colour.—Sheets printed neither too black nor too light.

Guide.—A piece of metal frequently used to denote the last line set.

Gutter-sticks.—Furniture used in imposition to separate the pages.

Half press.—When but one person works at the press.

Half-title.—The title of a book inserted in the upper portion of the first page of matter.

Head-sticks.—Furniture put at the head of pages in imposition, to make margin.

Hell.—The receptacle for broken or battered letters; the old-metal box; the shoe.

High-line.—Term applied to a type that ranges above the rest in a line.

High (*or low*) *to paper.*—Applied to a type cast higher or lower than the rest of the fount.

Horse.—The stage on the bank on which pressmen set the heap of paper.

Horsing.—Charging for work before it is executed.

Imposing.—Arranging and locking up a form of type in a chase.

Imposing-stone.—The stone on which compositors impose and correct forms.

Imprint.—The name of the printer or of the publisher appended to jobs or title-pages.

Inferior letters.—Small letters cast near the bottom of the line.

Inset.—Same as offcut.

Jeff.—To throw for a choice with quadrates instead of dice.

Justifying.—Spacing out lines accurately.

Keep in.—To crowd in by thin spacing.

Keep out.—To drive out or expand matter by wide spacing.

Kerned letter.—Type of which a part of the face hangs over the body.

Laying cases.—Filling cases with a fount of new type.

Laying pages.—Placing pages on the stone in a proper order for imposition.

Leaders.—Dots or hyphens placed at intervals of one or more ems in length, to guide the eye across the line to the folio in tables of contents, &c.

Leads.—Thin strips of metal cast of various thicknesses, quadrate-high, to separate lines of type.

Lean.—Close and solid matter.

Lean face.—Light, thin type.

Letter hangs.—When the page is out of square.

Letter-press printing.—Printing from types.

Ligatures.—Two or more letters cast on the same shank, as ff, fi, fl, ffi, ffl, æ, œ.

Locking up.—Tightening up a form by means of quoins.

Logotypes.—The same as ligatures.

Long cross.—The bar that divides a chase the longest way.

Long pull.—When the bar is brought close to the cheek of a press.

Low case.—When the compositor has set almost all the letters out of his case.

Lower case.—The case containing the small letters of the alphabet, figures, points, &c.

Low-line.—Applied to a type that ranges lower than the rest in a line.

Mackle.—When part of the impression appears double.

Make-up.—To arrange the lines of matter into pages.

Making margin.—In imposition, arranging the space between the pages of a form so that the margin will be properly proportioned.

Making ready.—Preparing a form on the press for printing.

Mallet.—A wooden hammer.

Matter.—Composed type.

Measure.—The width of a page.

Monk.—A black spot in a printed sheet, owing to the ink not being properly distributed.

Naked form.—A form without furniture.

Nicks.—Hollows cast in the front of the lower part of the shank of a type, to show the compositor how to place it in his stick.

Octavo, or *8vo.*—Eight pages to a form.

Odd page or folio.—The 1st, 3d, and all uneven-numbered pages.

Off.—Signifies that the pressman has worked off the form.

Offcut.—A portion of a sheet that is cut off before folding.

Off its feet.—When matter does not stand upright.

Open matter.—Matter widely leaded or containing numerous break-lines.

Out.—An omission marked in a proof by the reader.

Out of register.—When the pages do not back each other.

Overlay.—A scrap of paper pasted on the tympan-sheet to bring up the impression.

Overrunning.—Carrying words backward or forward in correcting.

Page-cord.—Twine used for tying up pages.

Passing the make-up.—Passing to the next hand in order the lines remaining (if any) after a compositor has made up his matter, together with the gauge and proper folio.

Peel.—A broad, thin board with a long handle.

Perfecting.—Printing the second form of a sheet.

Pi.—Type promiscuously intermingled.

Pick.—A particle of ink or paper imbedded in the hollow of a letter, filling up its face and occasioning a spot.

Pigs.—An ancient nickname given in derision by compositors to pressmen. The press-room was called a pigsty.

Planer.—A smooth block of wood used for levelling the surface of pages of type when imposed.

Planing down.—To bring down types evenly on their feet, by laying a planer on the page and striking it firmly with a mallet.

Platen.—The part of a printing-press which, under the influence of the lever, gives the impression to a sheet.

Point-holes.—Fine holes made by the points to register the second impression by.

Points.—Two thin pieces of steel with a point at one end, adjusted to the tympan with screws, to make register.

Quadrate.—A low square blank type, used to indent the first line of a paragraph, and to fill up blank spaces.

Quarters.—Octavos and twelves are said to be imposed in quarters, not from their equal divisions, but because they are imposed and locked up in four parts.

Quarto, or 4to.—Four pages to a form.

Quire.—Twenty-four sheets of paper.

Quoins.—Small wedges for locking up a form.

Quotation furniture.—Quotations cast of various sizes in length and width, to be used for blanking and as furniture.

Quotations.—Large hollowed quadrates.

Rack.—Receptacle for cases.

Ratting.—Working at less than the established prices.

Ream.—Twenty quires of paper.

Recto.—Right-hand page.

References.—Letters or characters serving to direct the reader's attention to notes at the foot of a page.

Register.—To cause the pages in a sheet to print precisely back to back.

Register sheet.—The sheet used to make register.

Reglet.—Thin furniture, of an equal thickness all its length. It is made to the thickness of type.

Reiteration.—The form printed on the second side.

Revise.—The last proof of a form before working it off.

Riding.—One colour falling on another. Type at the end of a line catching against a lead.

Rise.—A form is said to rise when, in raising it from the correcting-stone, no letters drop out.

Roller.—A hollow wooden cylinder covered with composition, which, set in an iron frame, revolves upon a rod, and is used for inking type.

Rounce.—The handle for running in and out the carriage of a hand-press.

Round pick.—A dot in a letter in a stereotype plate caused by an air-bubble.

Running title.—The title of the book or subject placed at the top of the pages.

Runs on sorts.—Requiring an inordinate proportion of particular letters.

Saw-block.—A box similar to a carpenter's mitre-block, to guide in cutting furniture, &c.

Schedule.—A list passed with the make-up, containing folios on which the compositor marks his name opposite to the pages set by him.

Set off.—When sheets that are newly worked off soil those that come in contact with them, they are said to set off.

Shank.—The square metal upon which the face of a letter stands.

Sheep's-foot.—An iron hammer with a claw-end.

Shooting-stick.—A wedge-shaped instrument for locking up a form.

Short cross.—The short bar which, crossing the long bar, divides the chase into quarters.

Shoulder.—The upper surface of the shank of a type not covered by the letter.

Side-sorts.—Types in the side and upper boxes of a case, consisting of letters not frequently used.

Side-sticks.—Sloping furniture on the outside of the pages next to the chase, where the quoins are inserted.

Signature.—A letter or a figure used at the bottom of the first page of a sheet, to direct the binder in placing the sheets in a volume.

Slice galley.—A galley with an upper false bottom, called a slice, used for large pages and jobs.

Slug.—A thick lead.

Slur.—A blurred impression in a printed sheet.

Solid pick.—A letter in a stereotype plate filled up with metal, resulting from an imperfect mould.

Sorts.—The letters in the several case-boxes are separately called sorts, in printers' and founders' language.

Space-rules.—Fine lines, cast type-high, and of even ems in length, for table and algebraical work.

Spaces.—Low blank types used to separate words.

Squabble.—A page or form is squabbled when the letters are twisted out of a square position.

Stand.—The frame on which the cases are placed.

Stem.—The straight flat strokes of a straight letter.

Stereotype printing.—Printing from plates.

Stet.—Written opposite to a word, to signify that the word erroneously struck out in a proof shall remain.

Sub.—A compositor occasionally employed on a daily paper, to fill the place of an absentee.

Superior letters.—Letters of a small face, cast by the founder near the top of the line.

Table-work.—Matter consisting partly of rules and figures.

Take, or *Taking.*—A given portion of copy.

Token.—Two hundred and fifty sheets.

Turn for a letter.—When a sort runs short, a letter of the same thickness is substituted, placed bottom upward.

Tympan.—A frame covered with parchment and attached to the press-bed, to lay the sheet on before printing.

Underlay.—A piece of paper or card placed under types or cuts to improve the impression.

Upper case.—The case containing capital and small capital letters, fractions, &c.

Verso.—Left-hand page.

Wayz-goose.—A term given in England to the annual dinner customary among printers there during the summer months.

White line.—A line of quadrates.

White page.—A blank page.

White paper.—Until the second side of a sheet is printed, pressmen call the heap white paper.

Working in pocket.—When the hands share equally their earnings on a work.

ABBREVIATIONS.

A.—Acting.
A. or Ans.—Answer.
A.A.G.—Assistant Adjutant-General.
A.A.P.S.—American Association for the Promotion of Science.
A.A.S.—*Academiæ Americanæ Socius*, Fellow of the American Academy (of Arts and Sciences).
A.A.S.S.—*Americanæ Antiquarianæ Societatis Socius*, Member of the American Antiquarian Society.
A.B.—*Artium Baccalaureus*, Bachelor of Arts.
A.B.C.F.M.—American Board of Commissioners for Foreign Missions.
Abp.—Archbishop.
Abr.—Abridgment.
A.C.—*Ante Christum*, before the birth of Christ.
A.C.—Archchancellor.
Acct.—Account.
A.C.S.—American Colonization Society.
A.D.—*Anno Domini*, in the year of our Lord.
A.D.C.—Aide-de-camp.
Adj.—Adjective.
Adjt.—Adjutant.
Adjt.-Gen.—Adjutant-General.
Ad lib.—*Ad libitum*, at pleasure.
Adm.—Admiral; Admiralty.
Adm. Co.—Admiralty Court.
Admr.—Administrator.
Admx.—Administratrix.
Ad v.—*Ad valorem*, at (or on) the value.
Adv.—Adverb.
Æt.—*Ætatis*, of age; aged.
A.F.B.S.—American and Foreign Bible Society.
A.G.—Adjutant-General.
Ag.—Argentum (silver).
Agr.—Agriculture.
A.G.S.S.—American Geographical and Statistical Society.
Agt.—Agent.
A.H.—*Anno Hegiræ*, in the year of the Hegira.
A.H.M.S.—American Home Missionary Society.
Ala.—Alabama.
Ald.—Alderman.
Alex.—Alexander.
Alg.—Algebra.
Alt.—Altitude.
A.M.—*Anno mundi*, in the year of the world.
A.M.—*Artium Magister*, Master of Arts.
A.M.—*Ante meridiem*, before noon; morning.
Amb.—Ambassador.
Amer.—American.
AMM.—*Amalgama*, amalgamation.
Amt.—Amount.
An.—*Anno*, in the year.
An. A.C.—*Anno ante Christum*, in the year before Christ.
Anat.—Anatomy.
Anc.—Ancient; anciently.
And.—Andrew.
Ang.-Sax.—Anglo-Saxon.
Anon.—Anonymous.
Anth.—Anthony.
Aor. or aor.—Aorist.
A.O.S.S.—*Americanæ Orientalis Societatis Socius*, Member of the American Oriental Society.
Ap.—Apostle; Appius.

Ap.—*Apud*, in the writings of; as quoted by.
A. P. G. or Ast. P. G.—Professor of Astronomy in Gresham College.
Apo.—Apogee.
Apoc.—Apocalypse.
App.—Appendix.
Apr.—April.
A. Q. M. G.—Assistant Quartermaster-General.
A. R.—*Anna Regina*, Queen Anne.
A. R.—*Anno regni*, year of the reign.
A. R. A.—Associate of the Royal Academy.
Arch.—Archibald.
Arg.—*Argumento*, by an argument drawn from such a law.
Arith.—Arithmetic.
Ark.—Arkansas.
A. R. R.—*Anno regni regis*, in the year of the reign of the king.
A. R. S. S.—*Antiquariorum Regiæ Societatis Socius*, Fellow of the Royal Society of Antiquaries.
Art.—Article.
A. S. or Assist. Sec.—Assistant Secretary.
A. S. A.—American Statistical Association.
A. S. S. U.—American Sunday-School Union.
Astrol.—Astrology.
Astron.—Astronomy.
A. T.—Archtreasurer.
A. T. S.—American Tract Society.
Ats.—At suit of.
Atty.—Attorney.
Atty.-Gen.—Attorney-General.
A. U. A.—American Unitarian Association.
Aub. Theol. Sem.—Auburn Theological Seminary.
A. U. C.—*Anno urbis conditæ*, or, *ab urbe conditâ*, in the year from the building of the city (Rome).
Aug.—August.
Aur.—*Aurum*, gold.
Auth. Ver.—Authorized Version (of the Bible).
Av.—Average; Avenue.
Avoir.—Avoirdupois.
A. Y. M.—Ancient York Masons.
b.—Born.
B. A.—Bachelor of Arts.
Bal.—Balance.
Balt.—Baltimore.
Bar.—Baruch.
Bart. or Bt.—Baronet.
Bbl.—Barrel.
B. C.—Before Christ.
B. C. L.—Bachelor of Civil Law.
B. D.—*Baccalaureus Divinitatis*, Bachelor of Divinity.
Bds. or bds.—Boards (bound in).
Benj.—Benjamin.
Bk.—Book.
B. LL.—*Baccalaureus Legum*, Bachelor of Laws.
B. M.—*Baccalaureus Medicinæ*, Bachelor of Medicine.
Bost.—Boston.
Bot.—Botany.
Bp.—Bishop.
B. R.—*Banco Regis* or *Reginæ*, the King's or Queen's Bench.
Br.—Brother.
Brig.—Brigade; Brigadier.
Brig.-Gen.—Brigadier-General.
Brit. Mus.—British Museum.
Bro.—Brother.
Br. Univ.—Brown University.
B. S.—Bachelor in the Sciences.
B. V.—*Beata Virgo*, Blessed Virgin.
B. V.—*Bene vale*, farewell.
C., Ch. or Chap.—Chapter.
C. or Cent.—*Centum*, a hundred.
cæt. par.—*Cæteris paribus*, other things being equal.
Cal.—California; Calends.
Can.—Canon.
Cant.—Canticles.
Cap. or c.—*Caput, capitulum*, chapter.
Caps.—Capitals.
Capt.—Captain.
Capt.-Gen.—Captain-General.
Cash.—Cashier.
ca. resp.—*Capias ad respondendum*, a legal writ.
ca. sa.—*Capias ad satisfaciendum*, a legal writ.
Cath.—Catherine.
C. B.—Companion of the Bath.
C. B.—*Communis Bancus*, Common Bench.
C. C.—Caius College; Account Current.
C. C. C.—Corpus Christi College.
C. C. P.—Court of Common Pleas.
C. E.—Canada East.

C. E.—Civil Engineer.
Cel. or Celt.—Celtic.
Cf. or cf.—*Confer*, compare.
C. G.—Commissary-General; Consul-General.
C. H.—Court-house.
Ch.—Church; Chapter; Charles.
Chanc.—Chancellor.
Chap.—Chapter
Chas.—Charles.
Chem.—Chemistry.
Chr.—Christopher.
Chron.—Chronicles.
Cin.—Cincinnati.
C. J.—Chief-Justice.
Clk.—Clerk.
C. M.—Common Metre.
C. M. G.—Companion of the Order of St. Michael and St. George.
Co.—Company; county.
C. O. D.—Cash (or collect) on delivery.
Col.—Colorado; Colonel; Colossians.
Coll.—Collector; Colloquial; College; Collection.
Com.—Commerce; Committee; Commentary; Commissioner; Commodore.
Com. Arr.—Committee of Arrangements.
Comdg.—Commanding.
Comm.—Commentary.
Comp.—Compare; Compound.
Com. Ver.—Common Version (of the Bible).
Con.—*Contra*, against; in opposition.
Con. Cr.—Contra, credit.
Conch.—Conchology.
Cong.—Congress.
Conj. or conj.—Conjunction.
Conn. or Ct.—Connecticut.
Const.—Constable; Constitution.
Cont.—Contra.
Cor.—Corinthians.
Corol.—Corollary.
Cor. Sec.—Corresponding Secretary.
C. P.—Common Pleas.
C. P.—Court of Probate.
C. P. S.—*Custos Privati Sigilli*, Keeper of the Privy Seal.
C. R.—*Custos Rotulorum*, Keeper of the Rolls.
Cr.—Creditor; credit.
Crim. Con.—Criminal conversation; adultery.
C. S.—Court of Sessions.
C. S.—*Custos Sigilli*, Keeper of the Seal.
Ct., cts.—Cent; Cents.
C. Theod.—*Codice Theodosiano*, in the Theodosian Code.
C. W.—Canada West.
Cwt.—Hundredweight.
Cyc.—Cyclopedia.
d.—*Denarius* or *Denarii*, penny or pence.
d.—Died.
D.—Five hundred.
Dan.—Daniel; Danish.
D. B. or Domesd. B.—Domesday-Book.
D. C.—*Da Capo*, again.
D. C.—District of Columbia.
D. C. L.—Doctor of Civil Law.
D. D.—*Divinitatis Doctor*, Doctor of Divinity.
Dea.—Deacon.
Dec.—December; Declination.
Deg.—Degree or degrees.
Del.—Delaware; Delegate.
Del. or del.—*Delineavit*, he (or she) drew it.
Dep.—Deputy.
Dept.—Department.
Deut.—Deuteronomy.
D. F.—Dean of the Faculty.
Dft. or Deft.—Defendant.
D. G.—*Dei gratiâ*, by the grace of God.
D. G.—*Deo gratias*, thanks to God.
Diam.—Diameter.
Dict.—Dictator; Dictionary.
Dim.—Diminutive.
Disc.—Discount.
Diss.—Dissertation.
Dist.—District.
Dist.-Atty.—District-Attorney.
D. M.—Doctor of Music.
Do.—*Ditto*, the same.
Dols.—Dollars.
D. O. M.—*Deo optimo maximo*, to God, the best, the greatest.
Doz.—Dozen.
D. P.—Doctor of Philosophy.
Dr.—Debtor; Doctor.
D. S.—*Dal segno*, from the sign.
d. s. b.—*Debit sans breve.*
D. T.—*Doctor Theologiæ*, Doctor of Divinity.
D. V.—*Deo volente*, God willing.
Dwt.—Pennyweight.
E.—East.
ea.—Each.

E. by S.—East by South.
Eben.—Ebenezer.
Eccl.—Ecclesiastes.
Ecclus.—Ecclesiasticus.
Ed.—Editor; Edition.
Edm.—Edmund.
Edw.—Edward.
E. E.—Errors excepted.
e. g.—*Exempli gratiâ*, for example.
e. g.—*Ex grege*, among the rest.
E. I.—East Indies or East India.
Eliz.—Elizabeth.
E. lon.—East longitude.
Encyc.—Encyclopedia.
E. N. E.—East-Northeast.
Eng.—England; English.
Ent.—Entomology.
Env. Ext.—Envoy Extraordinary.
Ep.—Epistle.
Eph.—Ephesians; Ephraim.
Esd.—Esdras.
E. S. E.—East-Southeast.
Esq.—Esquire.
Esth.—Esther.
et al.—*Et alii*, and others.
et seq.—*Et sequentia*, and what follows.
etc. or &c.—*Et cæteri*, *et cæteræ*, *et cætera*, and others; and so forth.
Ex.—Example.
Ex.—Exodus.
Exc.—Excellency; exception.
Exch.—Exchequer.
Exec. Com.—Executive Committee.
Execx. Executrix.
Exr. or Exec.—Executor.
Ez.—Ezra.
Ezek.—Ezekiel.
E. & O. E.—Errors and omissions excepted.
Fahr.—Fahrenheit.
F. A. M.—Free and Accepted Masons.
Far. Farthing.
F. A. S.—Fellow of the Antiquarian Society.
fcap. or fcp.—Foolscap.
F. D.—*Fidei Defensor* or *Defensatrix*, Defender of the Faith.
Fe.—Ferrum (iron).
Feb.—February.
Fec. *Fecit*, he did it.
Fem.—Feminine.
F. E. S.—Fellow of the Entomological Society; of the Ethnological Society.
Ff.—The Pandects.
F. G. S.—Fellow of the Geological Society.
F. H. S.—Fellow of the Horticultural Society.
fi. fa.—*Fieri facias*, cause it to be done.
Fid. Def.—Defender of the Faith.
Fig.—Figure.
Fir.—Firkin.
Fla.—Florida.
F. L. S.—Fellow of the Linnæan Society.
Fol.—Folio.
For.—Foreign.
F. P. S.—Fellow of the Philological Society.
Fr.—Franc; francs.
Fr.—*Fragmentum*, fragment.
Fr.—Francis.
F. R. A. S.—Fellow of the Royal Astronomical Society.
F. R. C. S. L.—Fellow of the Royal College of Surgeons, London.
Fred.—Frederick.
F. R. G. S.—Fellow of the Royal Geographical Society.
Fri.—Friday.
F. R. S.—Fellow of the Royal Society.
Frs.—Frisian.
F. R. S. E.—Fellow of the Royal Society, Edinburgh.
F. R. S. L.—Fellow of the Royal Society, London.
F. R. S. L.—Fellow of the Royal Society of Literature.
F. S. A.—Fellow of the Society of Arts.
F. S. A. E.—Fellow of the Society of Antiquaries, Edinburgh.
Ft.—Foot; feet; Fort.
Fur.—Furlong.
F. Z. S.—Fellow of the Zoological Society.
G. or g.—Guineas.
G. A.—General Assembly.
Ga.—Georgia.
Gal.—Galatians; Gallon.
G. B.—Great Britain.
G. C.—Grand Chapter.
G. C. B.—Grand Cross of the Bath.
G. C. H.—Grand Cross of Hanover.
G. C. L. H.—Grand Cross of the Legion of Honour
G. E.—Grand Encampment.
Gen.—Genesis; General.
Gent.—Gentleman.
Geo.—George.

Geog.—Geography.
Geol.—Geology.
Geom.—Geometry.
Ger.—Germany; German.
G. L.—Grand Lodge.
Gl.—*Glossa*, a gloss.
G. M.—Grand Master.
G. O.—General Order.
Goth.—Gothic.
Gov.—Governor.
Gov.-Gen.—Governor-General.
G. R.—*Georgius Rex*, King George.
Gr.—Greek; Gross.
Gram.—Grammar.
Gro.—Gross.
Grot.—Grotius.
h. a.—*Hoc anno*, this year.
Hab.—Habakkuk.
Hab. corp.—*Habeas corpus*, you may have the body.
Hab. fa. poss.—*Habere facias possessionem.*
Hab. fa. seis.—*Habere facias seisinam.*
Hag.—Haggai.
Ham. Coll.—Hamilton College.
H. B. C.—Hudson's Bay Company.
H. B. M.—His or Her Britannic Majesty
H. C.—House of Commons.
Hdkf.—Handkerchief.
h. e.—*Hoc est*, that is, or this is.
Heb.—Hebrews.
Her.—Heraldry.
Hf.-bd.—Half-bound.
Hg.—*Hydrargyrum*, mercury.
Hhd.—Hogshead.
Hist.—History.
H. J. S.—*Hic jacet sepultus*, Here lies buried.
H. L.—House of Lords.
H. M.—His Majesty.
H. M. P.—*Hoc monumentum posuit*, erected this monument.
Hon.—Honourable.
Hort.—Horticulture.
Hos.—Hosea.
H. R.—House of Representatives.
H. R. E.—Holy Roman Emperor.
H. R. H.—His Royal Highness.
H. R. I. P.—*Hic requiescit in pace*, Here rests in peace.
H. S.—*Hic situs*, Here lies.
H. S. H.—His Serene Highness.
h. t.—*Hoc titulum*, this title; *hoc tituli*, in or under this title.
h. v.—*Hoc verbum*, this word; *his verbis*, in these words.
Hund.—Hundred.
I. II. III.—One, two, three, or first, econd, third.
Ia.—Iowa.
Ib. or ibid.—*Ibidem*, in the same place.
Ich.—Ichthyology.
Ictus.—*Jurisconsultus.*
Id.—*Idem*, the same.
Id. T.—Idaho Territory.
i. e.—*Id est*, that is.
I. H. M.—*Jesus hominum mundi*, Jesus the Saviour of the world.
I. H. S.—*Jesus hominum Salvator*, Jesus the Saviour of men.
ij.—Two (*med.*).
Ill.—Illinois.
In.—Inch; inches.
incog.—*Incognito*, unknown.
Incor.—Incorporated.
Ind.—Indiana; Index.
Ind. Ter.—Indian Territory.
Indef.—Indefinite.
Inf.—*Infra*, beneath or below.
in f.—*In fine*, at the end of the title, law, or paragraph quoted.
in lim.—*In limine*, at the outset.
in loc.—*In loco*, in the place; on the passage.
in pr.—*In principio*, in the beginning and before the first paragraph of a law.
I. N. R. I.—*Jesus Nazarenus, Rex Judæorum*, Jesus of Nazareth, King of the Jews.
Inst.—Instant, of this month; Institutes.
In sum.—*In summa*, in the summary.
Int.—Interest.
Interj.—Interjection.
in trans.—*In transitu*, on the passage.
Introd.—Introduction.
I. O. O. F.—Independent Order of Odd-Fellows.
I. O. U.—I owe you.
I. q.—*Idem quod*, the same as.
Isa.—Isaiah.
Isl.—Island.
Ital.—Italic; Italian.
IV.—Four or fourth.
IX.—Nine or ninth.
J.—Justice or Judge. JJ.—Justices.
j.—One (*med.*).
J. A.—Judge-Advocate.

Jac.—Jacob.
Jan.—January.
Jas.—James.
J. C. D.—*Juris Civilis Doctor*, Doctor of Civil Law.
J. D.—*Jurum Doctor*, Doctor of Laws.
Jer.—Jeremiah.
Jno.—John.
Jona.—Jonathan.
Jos.—Joseph.
Josh.—Joshua.
J. P.—Justice of the Peace.
J. Prob.—Judge of Probate.
J. R.—*Jacobus Rex*, King James.
Jr. or Jun.—Junior.
J. U. D. or J. V. D.—*Juris utriusque Doctor*, Doctor of both Laws (of the Canon and the Civil Law).
Jud.—Judith.
Judg.—Judges.
Judge-Adv.—Judge-Advocate.
Jul. Per.—Julian Period.
Jus. P.—Justice of the Peace.
Just.—Justinian.
J. W.—Junior Warden.
K.—King.
K. A.—Knight of St. Andrew, in Russia.
K. A. N.—Knight of Alexander Nevskoi, in Russia.
Kan.—Kansas.
K. B.—King's Bench.
K. B.—Knight of the Bath.
K. B. A.—Knight of St. Bento d'Avis, in Portugal.
K. B. E.—Knight of the Black Eagle, in Russia.
K. C.—King's Council.
K. C.—Knight of the Crescent, in Turkey.
K. C. B.—Knight Commander of the Bath.
K. C. H.—Knight Commander of Hanover.
K. C. S.—Knight of Charles III. of Spain.
K. E.—Knight of the Elephant, in Denmark.
K. F.—Knight of Ferdinand of Spain.
K. F. M.—Knight of St. Ferdinand and Merit, in Sicily.
K. G.—Knight of the Garter.
K. G. C.—Knight of the Grand Cross.
K. G. C. B.—Knight of the Grand Cross of the Bath.
K. G. F.—Knight of the Golden Fleece, in Spain.
K. G. H.—Knight of the Guelphs of Hanover.
K. G. V.—Knight of Gustavus Vasa of Sweden.
K. H.—Knight of Hanover.
Ki.—Kings.
Kingd.—Kingdom.
K. J.—Knight of St. Joachim.
K. L. or K. L. A.—Knight of Leopold of Austria.
K. L. H.—Knight of the Legion of Honour.
K. M.—Knight of Malta.
K. Mess.—King's Messenger.
K. M. H.—Knight of Merit, in Holstein.
K. M. J.—Knight of Maximilian Joseph of Bavaria.
K. M. T.—Knight of Maria Theresa of Austria.
Knick.—Knickerbocker.
K. N. S.—Knight of the Royal North Star, in Sweden.
Knt. or Kt.—Knight.
K. P.—Knight of St. Patrick.
K. R. C.—Knight of the Red Cross.
K. R. E.—Knight of the Red Eagle, in Prussia.
K. S.—Knight of the Sword, in Sweden.
K. S. A.—Knight of St. Anne of Russia.
K. S. E.—Knight of St. Esprit, in France.
K. S. F.—Knight of St. Fernando of Spain.
K. S. F. M.—Knight of St. Ferdinand and Merit, in Naples.
K. S. G.—Knight of St. George of Russia.
K. S. H.—Knight of St. Hubert of Bavaria.
K. S. J.—Knight of St. Januarius of Naples.
K. S. L.—Knight of the Sun and Lion, in Persia.
K. S. M. & S. G.—Knight of St. Michael and St. George of the Ionian Islands.
K. S. P.—Knight of St. Stanislaus of Poland.
K. S. S.—Knight of the Southern Star of the Brazils.
K. S. S.—Knight of the Sword, in Sweden.
K. S. W.—Knight of St. Wladimir of Russia.
K. T.—Knight of the Thistle; Knight Templar.
Kt.—Knight.

K.T.S.—Knight of the Tower and Sword, in Portugal.
K.W.—Knight of William of the Netherlands.
K.W.E.—Knight of the White Eagle, in Poland.
Ky.—Kentucky.
L.—Fifty or fiftieth.
L.—*Liber*, book.
L, £, or l.—*Libra* or *libræ*, pound or pounds sterling.
L. or £, s. d.—Pounds, shillings, pence.
La.—Louisiana.
Lam.—Lamentations.
Lat.—Latitude; Latin.
Lb. or ℔.—*Libra* or *libræ*, pound or pounds in weight.
L.C.—Lord Chancellor; Lord Chamberlain.
L.C.—Lower Canada.
Liv.—*Livre*, book.
l.c.—Lowercase.
L.C.J.—Lord Chief-Justice.
L.D.—Lady-Day.
Ld.—Lord.
Ldp.—Lordship.
Leg.—Legate.
Legis.—Legislature.
Lev.—Leviticus.
Lex.—Lexicon.
L.I.—Long Island.
Lib.—*Liber*, book.
Lieut.—Lieutenant.
Lieut.-Col.—Lieutenant-Colonel.
Lieut.-Gen.—Lieutenant-General.
Lieut.-Gov.—Lieutenant-Governor.
Linn.—Linnæan.
Lit.—Literally; Literature.
LL.B.—*Legum Baccalaureus*, Bachelor of Laws.
LL.D.—*Legum Doctor*, Doctor of Laws.
l.l.—*Loco laudato*, in the place quoted.
loc. cit.—*Loco citato*, in the place cited.
Lon.—Longitude.
L.S.—*Locus sigilli*, place of the seal.
Lt.—Lieutenant.
LX.—Sixty or sixtieth.
LXX.—Seventy or seventieth.
LXX.—The Septuagint (Version of the Old Testament).
LXXX.—Eighty or eightieth.
M.—*Meridies*, noon.
M.—*Mille*, a thousand.
M. or Mons.—Monsieur.
M.A.—Master of Arts.
Macc.—Maccabees.
Mad.—Madam.
Mad. Univ.—Madison University.
Maj.—Major.
Maj.-Gen.—Major-General.
Mal.—Malachi.
Man.—Manasses.
Mar.—March.
March.—Marchioness.
Marg.—Margin.
Marg. Tran.—Marginal Translation.
Marq.—Marquis.
Masc.—Masculine.
Mass.—Massachusetts.
Math.—Mathematics; Mathematician.
Matt.—Matthew.
Max.—Maxim.
M.B.—*Medicinæ Baccalaureus*, Bachelor of Medicine.
M.B.—*Musicæ Baccalaureus*, Bachelor of Music.
M.B.F. et H.—Great Britain, France, and Ireland.
M.C.—Member of Congress.
Mch.—March.
M.D.—*Medicinæ Doctor*, Doctor of Medicine.
Md.—Maryland.
Mdlle.—Mademoiselle.
M.E.—Methodist Episcopal; Military or Mechanical Engineer.
Me.—Maine.
Med.—Medicine.
Mem.—Memorandum.
Mem.—*Memento*, remember.
Merc.—Mercury.
Messrs. or MM.—Messieurs, Gentlemen.
Met.—Metaphysics.
Metal.—Metallurgy.
Meteor.—Meteorology.
Meth.—Methodist.
Mex.—Mexico or Mexican.
M.-Goth.—Mœso-Gothic.
M.H.S.—Massachusetts Historical Society.
M.H.S.—Member of the Historical Society.
Mic.—Micah.
Mich.—Michigan.
Mil.—Military.
Min.—Mineralogy.
Min.—Minute.
Minn.—Minnesota.

Min. Plen.—Minister Plenipotentiary.
Miss.—Mississippi.
M. L. A.—Mercantile-Library Association.
MM.—Their Majesties.
MM.—Messieurs, Gentlemen.
MM.—Two thousand.
Mme.—Madame.
M. M. S.—Moravian Missionary Society.
M. M. S. S.—*Massachusettensis Medicinæ Societatis Socius*, Fellow of the Massachusetts Medical Society.
Mo.—Missouri; Mouth.
Mod.—Modern.
Mon.—Monday.
Mons.—Monsieur, Sir.
Mos.—Months.
M. P.—Member of Parliament; Member of Police.
M. P. P.—Member of Provincial Parliament.
M. R.—Master of the Rolls.
Mr.—Mister.
M. R. A. S.—Member of the Royal Asiatic Society; Member of the Royal Academy of Science.
M. R. C. C.—Member of the Royal College of Chemistry.
M. R. C. S.—Member of the Royal College of Surgeons.
M. R. G. S.—Member of the Royal Geographical Society.
M. R. I.—Member of the Royal Institution.
M. R. I. A.—Member of the Royal Irish Academy.
Mrs.—Mistress.
M. R. S. L.—Member of the Royal Society of Literature.
M. S.—*Memoriæ sacrum*, Sacred to the memory.
M. S.—Master of the Sciences.
MS.—*Manuscriptum*, manuscript.
MSS.—Manuscripts.
Mt.—Mount or mountain.
Mus. B.—Bachelor of Music.
Mus. D.—Doctor of Music.
M. W.—Most Worthy; Most Worshipful.
Myth.—Mythology.
N.—North; Number; Noun; Neuter.
n.—Note.
N. A.—North America.
Nah.—Nahum.
Nat.—Natural.
Nat. Hist.—Natural History.
Nath.—Nathanael or Nathaniel.
N. B.—New Brunswick; North British.
N. B.—*Nota bene*, mark well; take notice.
N. C.—North Carolina.
N. E.—New England; Northeast.
Neb.—Nebraska.
Neh.—Nehemiah.
n. e. i.—*Non est inventus*, he is not found.
nem. con. or nem. diss.—*Nemine contradicente*, or *nemine dissentiente*, no one opposing; unanimously.
Neut.—Neuter (gender).
New M.—New Mexico.
New Test. or N. T.—New Testament.
N. F.—Newfoundland.
N. G.—New Granada; Noble Grand.
N. H.—New Hampshire; New Haven.
N. H. H. S.—New Hampshire Historical Society.
Ni. pri.—Nisi prius.
N. J.—New Jersey.
n. l.—*Non liquet*, it does not appear.
N. lat.—North latitude.
N. M.—New Mexico.
N. N. E.—North-northeast.
N. N. W.—North-northwest.
N. O.—New Orleans.
No.—*Numero*, number.
Nol pros.—*Nolle prosequi*, unwilling to proceed.
Nom. or nom.—Nominative.
Non con.—Not content; dissenting (House of Lords).
Non cul.—*Non culpabilis*, Not guilty.
Non obst.—*Non obstante*, notwithstanding.
Non pros.—*Non prosequitur*, he does not prosecute.
Non seq.—*Non sequitur*, it does not follow.
Nos.—Numbers.
Nov.—November.
N. P.—Notary Public.
N. S.—New Style (after 1752); Nova Scotia.
N. T.—New Testament; Nevada Territory.
N. u.—Name or names unknown.
Num.—Numbers; Numeral.
N. V. M.—Nativity of the Virgin Mary.
N. W.—Northwest.
N. Y.—New York.

N.Y.H.S.—New York Historical Society.
O.—Ohio.
Ob.—*Obiit*, he or she died.
Obad.—Obadiah.
Obs.—Obsolete; Observatory; Observation.
Obt. or obdt.—Obedient.
Oct.—October.
O.F.—Odd-Fellow or Odd-Fellows.
Old Test. or O.T.—Old Testament.
Olym.—Olympiad.
Opt.—Optics.
Or.—Oregon.
Orig.—Originally.
Ornith.—Ornithology.
O.S.—Old Style (before 1752).
O.T.—Oregon Territory; Old Testament.
O.U.A.—Order of United Americans.
Oxf.—Oxford.
Oxon.—*Oxonia*, *Oxonii*, Oxford.
Oz.—Ounce.
P.—*Pondere*, by weight.
P. or p.—Page; Part; Participle.
Pa. or Penn.—Pennsylvania.
Pal.—Palæontology.
Par.—Paragraph.
Par. Pas.—Parallel passage.
Parl.—Parliament.
Pathol.—Pathology.
Payt.—Payment.
Pb.—*Plumbum*, lead.
P.B.—*Philosophiæ Baccalaureus*, Bachelor of Philosophy.
P.C.—*Patres Conscripti*, Conscript Fathers; Senators.
P.C.—Privy Council; Privy Councillor.
P.D.—*Philosophiæ Doctor*, Doctor of Philosophy.
Pd.—Paid.
P.E.—Protestant Episcopal.
P.E.I.—Prince Edward Island.
Penn.—Pennsylvania.
Pent.—Pentecost.
Per, or pr.—By the, or per lb.
Per an.—*Per annum*, by the year.
Per cent.—*Per centum*, by the hundred.
Peri.—Perigee.
Pet.—Peter.
P.G.—Past Grand.
Phar.—Pharmacy.
Ph. B.—*Philosophiæ Baccalaureus*, Bachelor of Philosophy.
Ph. D.—*Philosophiæ Doctor*, Doctor of Philosophy.
Phil.—Philip; Philippians; Philosophy; Philemon.
Phila. or Phil.—Philadelphia.
Philem.—Philemon.
Philom.—*Philomathes*, a lover of learning.
Philomath.—*Philomathematicus*, a lover of the mathematics.
Phil. Trans.—Philosophical Transactions.
Phren.—Phrenology.
P.H.S.—Pennsylvania Historical Society.
Pinx. or pxt.—*Pinxit*, he (or she) painted it.
Pl. or Plur.—Plural.
Plff.—Plaintiff.
P.M.—*Post meridiem*, afternoon, evening.
P.M.—Postmaster; Passed Midshipman.
P.M.G.—Postmaster-General; Professor of Music in Gresham College.
P.O.—Post-Office.
Poet.—Poetical.
Pop.—Population.
Port.—Portugal or Portuguese.
P.P.—*Pater Patriæ*, the father of his country.
P.P.C.—*Pour prendre congé*, to take leave.
Pp. or pp.—Pages.
Pph.—Pamphlet.
P.R.—*Populus Romanus*, the Roman people.
P.R.A.—President of the Royal Academy.
P.R.C.—*Post Romanum conditum*, from the building of Rome.
Pref.—Preface.
Prep.—Preposition.
Pres.—President.
Prin.—Principally.
Prob.—Problem.
Prof.—Professor.
Pron.—Pronoun; Pronunciation.
Prop.—Proposition.
Prot.—Protestant.
Pro tem.—*Pro tempore*, for the time being.
Prov.—Proverbs; Provost.
prox.—*Proximo*, next (month).

P. R. S.—President of the Royal Society.
P. S.—*Post scriptum*, Postscript.
P. S.—Privy Seal.
Ps.—Psalm or Psalms.
Pt.—Part; Pint; Payment; Point; Port.
P. t.—Post-town.
P. Th. G.—Professor of Theology in Gresham College.
Pub.—Publisher; Publication; Published; Public.
Pub. Doc.—Public Documents.
P. v.—Post-village.
Pwt.—Pennyweight; pennyweights.
Pxt.—*Pinxit*, he (or she) painted it.
Q.—Queen.
Q.—Question.
q.—*Quasi*, as it were; almost.
Q. B.—Queen's Bench.
Q. C.—Queen's College.
Q. C.—Queen's Counsel.
q. d.—*Quasi dicat*, as if he should say; *quasi dictum*, as if said; *quasi dixisset*, as if he had said.
q. e.—*Quod est*, which is.
q. e. d.—*Quod erat demonstrandum*, which was to be proved.
q. e. f.—*Quod erat faciendum*, which was to be done.
q. e. i.—*Quod erat inveniendum*, which was to be found out.
q. l.—*Quantum libet*, as much as you please.
Q. M.—Quartermaster.
qm.—*Quomodo*, how; by what means.
Q. M. G.—Quartermaster-General.
q. p. or q. pl.—*Quantum placet*, as much as you please.
Qr.—Quarter.
Q. S.—Quarter Sessions.
q. s.—*Quantum sufficit*, a sufficient quantity.
Qt.—Quart.
qu. or qy.—*Quære*, inquire; query.
Quar.—Quarterly.
Ques.—Question.
q. v.—*Quod vide*, which see; *quantum vis*, as much as you will.
℞.—*Recipe*, take.
R.—*Regina*, Queen; *Rex*, King.
R.—River; Rood; Rod.
R. A.—Royal Academy; Royal Academician.
R. A.—Royal Arch.
R. A.—Royal Artillery.
RC.—*Rescriptum*, a counterpart.
R. E.—Royal Engineers.
Rec.—Recipe or Recorder.
Recd.—Received.
Rec. Sec.—Recording Secretary.
Rect.—Rector; Receipt.
Ref.—Reference.
Ref. Ch.—Reformed Church.
Reg.—Register; Regular.
Reg. Prof.—Regius Professor.
Regr.—Registrar.
Regt.—Regiment.
Rel.—Religion.
Rep.—Representative; Reporter.
Rev.—Reverend; Revelation (Book of); Review; Revenue; Revise.
Rhet.—Rhetoric.
R. I.—Rhode Island.
Richd.—Richard.
R. I. H. S.—Rhode Island Historical Society.
R. M.—Royal Marines; Royal Mail.
R. M. S.—Royal Mail Steamer.
R. N.—Royal Navy.
R. N. O.—*Riddare af Nordstjerne*, Knight of the Order of the Polar Star.
Ro.—*Recto*, right-hand page.
Robt.—Robert.
Rom.—Romans (Book of).
Rom. Cath.—Roman Catholic.
R. P.—*Regius Professor*, the King's Professor.
R. R.—Railroad.
R. S.—Recording Secretary.
Rs.—*Responsus*, to answer.
R. S. A.—Royal Society of Antiquaries; Royal Scottish Academy.
R. S. D.—Royal Society of Dublin.
R. S. E.—Royal Society of Edinburgh.
R. S. L.—Royal Society of London.
R. S. S.—*Regiæ Societatis Socius*, Fellow of the Royal Society
Rt. Hon.—Right Honourable.
Rt. Rev.—Right Reverend.
Rt. Wpful.—Right Worshipful.
R. W.—Right Worthy.
R. W. O.—*Riddare af Wasa Orden*, Knight of the Order of Wasa.
S.—South; Saint; Scribe; Sulphur; Sunday; Sun; Series.
S.—*Solidus*, a shilling.
S. A.—South America; South Africa; South Australia.

s. a.—*Secundum artem*, according to art.
Sam.—Samuel.
Sansc.—Sanscrit.
S. A. S.—*Societatis Antiquariorum Socius*, Fellow of the Society of Antiquarians.
Sat.—Saturday.
Sax.—Saxon.
Sax. Chron.—Saxon Chronicle.
S. C.—*Senatûs Consultum*, a decree of the Senate; South Carolina.
Sc.—*Sculpsit*, he (or she) engraved it.
sc. or scil.—*Scilicet*, namely.
Scan. Mag.—*Scandalum magnatum*, great scandal.
S. caps.—Small capitals.
Schol.—*Scholium*, a note.
Schr.—Schooner.
Sci. fa.—*Scire facias.*
Sclav.—Sclavonic.
Sculp. or sculp.—*Sculpsit*, he (or she) engraved it.
S. D.—*Salutem dicit*, sends health.
S. E.—Southeast.
Sec.—Secretary; Second.
Sec. Leg.—Secretary of Legation.
Sec. leg.—*Secundum legem*, according to law.
Sec. reg.—*Secundum regulam*, according to rule.
Sect.—Section.
Sem.—*Semble*, it seems.
Sen.—Senate; Senator; Senior.
Sept.—September; Septuagint.
Seq.—*Sequentia*, following; *sequitur*, it follows.
Ser.—Series.
Serg.—Sergeant.
Serg.-Maj.—Sergeant-Major.
Servt.—Servant.
S. G.—South Georgia; Solicitor-General.
Shak.—Shakspeare.
S. H. S.—*Societatis Historiæ Socius*, Fellow of the Historical Society
Sing.—Singular.
S. Isl.—Sandwich Islands.
S. J.—Society of Jesus.
S. J C.—Supreme Judicial Court.
Skr.—Sanscrit.
S. L.—Solicitor at Law (Scot.).
S. lat.—South latitude.
S. M.—State Militia; Short Metre; Sergeant-Major; Sons of Malta.
S. M. Lond. Soc. Cor.—*Societatis Medicæ Londonensis Socius Cor.*, Corresponding Member of the London Medical Society.
s. n.—*Secundum naturam*, according to nature.
Soc. Isl.—Society Islands.
Sol.—Solomon; Solution.
Sol.-Gen.—Solicitor-General.
S. of Sol.—Song of Solomon.
S. P.—*Sine prole*, without issue.
S. P. A. S.—*Societatis Philosophicæ Americanæ Socius*, Member of the American Philosophical Society.
S. P. G.—Society for the Propagation of the Gospel.
Sp. gr.—Specific gravity.
S. P. Q. R.—*Senatus Populusque Romani*, the Senate and people of Rome.
Sq. ft.—Square foot or square feet.
Sq. in.—Square inch or inches.
Sq. m.—Square mile or miles.
Sq. r.—Square rood or roods.
Sq. yd.—Square yard.
Sr.—Sir; Senior.
S. R. I.—*Sacrum Romanum Imperium*, Holy Roman Empire.
S. R. S.—*Societatis Regiæ Socius*, Fellow of the Royal Society.
S. S.—Sunday-school.
SS.—Saints.
SS. or ss.—*Scilicet*, to wit.
ss.—*Semis*, half.
S. S. C.—Solicitor before the Supreme Court (Scotland).
S. S. E.—South-southeast.
S. S. W.—South-southwest.
St.—Saint; Street; Strait.
Stat.—Statute.
S. T. D.—*Sacræ Theologiæ Doctor*, Doctor of Divinity.
Ster. or Stg.—Sterling.
S. T. P.—*Sacræ Theologiæ Professor*, Professor of Divinity.
Su.—Sunday
Subj.—Subjunctive.
Subst.—Substantive.
Su.-Goth.—Suio-Gothic.
Sun. or Sund.—Sunday.
Sup.—Supplement; Superfine.
Supt.—Superintendent.
Surg.—Surgeon; Surgery
Surg.-Gen.—Surgeon-General.
Surv.—Surveyor.
Surv.-Gen.—Surveyor-General.

Sus.—Susannah.
s.v.—*Sub verbo*, under the word or title.
S.W.—Southwest.
Syn.—Synonym; Synonymous.
T.—Territory.
T.—*Tutti*, all together.
T. or tom.—Tome, volume.
Ta.—Tantalum (Columbium).
T.E.—Topographical Engineers.
Tenn.—Tennessee.
Ter.—Territory.
Tex.—Texas.
Text. Rec.—*Textus Receptus*, the Received Text.
Th. or Thurs.—Thursday.
Theo.—Theodore.
Theol.—Theology; Theological.
Theoph.—Theophilus.
Thess.—Thessalonians.
Tho'.—Though.
Thos.—Thomas.
Thro'.—Through.
Tim.—Timothy.
Tit.—Titus.
T.O.—Turn over.
Tob.—Tobit.
Tom.—Volume.
Topog.—Topography; Topographical.
Tr.—Transpose; Translator; Translation.
Tr.—Trustee. Trs.—Trustees.
tr.—*Trillo*, a shake.
Trans.—Translator; Translation; Transactions.
Treas.—Treasurer.
Trin.—Trinity.
Tues. or Tu.—Tuesday.
Typ.—Typographer.
U.C.—Upper Canada.
U.C.—*Urbe condita*, year of Rome.
U.E.I.C.—United East India Company.
U.G.R.R.—Underground Railway.
U.J.C.—*Utriusque Juris Doctor*, Doctor of both Laws.
U.K.—United Kingdom.
ult.—*Ultimo*, last; of the last month.
Unit.—Unitarian.
Univ.—University.
U.S.—United States.
u.s.—*Ut supra* or *uti supra*, as above.
U.S.A.—United States Army.
U.S.A.—United States of America.
U.S.M.—United States Mail.
U.S.M.—United States Marines.
U.S.M.A.—United States Military Academy.
U.S.N.—United States Navy.
U.S.N.A.—United States Naval Academy.
U.S.S.—United States Senate.
U.T.—Utah Territory.
V.—Five or fifth.
V.—Violin. VV.—Violins.
v. or vid.—*Vide*, see.
v. or vs.—*Versus*, against; *Versiculo*, in such a verse.
Va.—Virginia.
Vat.—Vatican.
V.C.—Vice-Chancellor.
V.D.M.—*Verbi Dei Minister*, Minister of God's Word.
Ven.—Venerable.
Ver.—Verse.
V.G.—Vicar-General.
v.g.—*Verbi gratiâ*, as for example.
VI.—Six or sixth.
VII.—Seven or seventh.
VIII.—Eight or eighth.
Vice-Pres. or V.P.—Vice-President.
Visc.—Viscount.
viz. or vi.—*Videlicet*, to wit; namely; that is to say.
Vo.—*Verso*, left-hand page.
Vol.—Volume.
V.R.—*Victoria Regina*, Queen Victoria.
Vt.—Vermont.
Vul.—Vulgate (Version).
W.—West.
Wash.—Washington.
Wed.—Wednesday.
West. Res. Coll.—Western Reserve College.
w.f.—Wrong fount.
Whf.—Wharf.
W.I.—West India.
Wisc.—Wisconsin.
Wisd.—Wisdom (Book of).
Wk.—Week.
W. lon.—West longitude.
W.M.—Worshipful Master.
Wm.—William.
W.M.S.—Wesleyan Missionary Society.
W.N.W.—West-northwest.
Wpful.—Worshipful.
W.S.—Writer to the Signet.
W.S.W.—West-southwest.
W.T.—Washington Territory.
Wt.—Weight.

W. Va.—West Virginia.
X.—Ten or tenth.
XI.—Eleven.
XII.—Twelve.
XIII.—Thirteen.
XIV.—Fourteen.
XV.—Fifteen.
XVI.—Sixteen.
XVII.—Seventeen.
XVIII.—Eighteen.
XIX.—Nineteen.
XX.—Twenty.
XXX.—Thirty.
XL.—Forty.
XC.—Ninety.
X. or Xt.—Christ.
Xmas or Xm.—Christmas.
Xn. or Xtian.—Christian.
Xnty. or Xty.—Christianity.
Xper or Xr. Christopher.
Yd.—Yard.
y. or y^{e}—The.
y^{m}—Them.
y^{n}—Then.
y^{r}—Their; Your.
y^{s}—This.
y^{t}—That.
Y. M. C. A.—Young Men's Christian Association.
Yrs.—Years; Yours.
Zach.—Zachary.
Zech.—Zechariah.
Zeph.—Zephaniah.
Zool.—Zoology.
&.—And.

BRONSTRUP PRESS.

FOREIGN WORDS AND PHRASES,

WITH TRANSLATIONS.

A bas.—Down with.
A capite ad calcem.—From head to foot.
A fin.—To the end.
A fortiori.—With stronger reason.
A l'abandon.—At random.
A la bonne heure.—Luckily; in good time.
A la dérobée.—By stealth.
A la mode.—According to the fashion.
A main armée.—With force of arms.
A mensa et thoro.—From bed and board.
A posteriori.—From effect to cause; from the latter.
A priori.—From cause to effect; from the former.
A vinculo matrimonii.—From the tie of marriage.
A votre santé.—To your health.
Ab extra.—From without.
Ab initio.—From the beginning.
Ab origine.—From the beginning.
Ab ovo.—From the beginning.
Ab urbe conditâ.—From the building of the city (Rome); abridged A. U. C.
Absit invidia.—All offence apart; let there be no malice.
Absit omen.—May it not prove ominous.
Ac etiam.—And also.
Actum est de republica.—It is all over with the commonwealth.
Ad absurdum.—To show the absurdity.
Ad arbitrium.—At pleasure.
Ad astra per aspera.—To the stars through difficulties.
Ad captandum vulgus.—To catch the mob or the vulgar.
Ad eundem.—To the same point or degree.
Ad finem.—To the end.
Ad Græcas Calendas.—An indefinite postponement. (The Greeks had no calends.)
Ad hominem.—To the man (that is, to the interests or the passions of the man).
Ad infinitum.—Without end.
Ad inquirendum.—For inquiry.
Ad interim.—In the mean while.
Ad libitum.—At pleasure.
Ad litem.—For the action (at law).
Ad nauseam.—To a disgusting degree.
Ad referendum.—For further consideration.
Ad rem.—To the purpose.
Ad unguem.—To the nail; exactly; nicely.
Ad valorem.—According to the value.
Addendum.—An addition or appendix.
Adhuc sub judice lis est.—The affair is not yet decided.
Ægrescit medendo.—The remedy is worse than the disease.
Æquam servare mentem.—To preserve an equable mind.
Æquo animo.—With an equable mind.
Ære perennius.—More lasting than brass; enduring ever.
Affaire du cœur.—A love-affair; an amour.
Afflatus.—Inspiration.
Agenda.—Things to be done.
Aide-de-camp.—Assistant to a general.

Aide-toi, et le ciel t'aidera.—Help thyself, and Heaven will help thee.
Alere flammam.—To feed the flame
Al fresco.—In the open air.
Alga.—A kind of sea-weed.
Alguazil.—A Spanish constable.
Alias.—Otherwise; elsewhere.
Alibi.—Elsewhere; not present.
Alis volat propriis.—She flies with her own wings.
Aliunde.—From some other quarter or person.
Allemande.—A kind of German dance.
Alma mater.—Benign mother (applied to a university).
Alter ego.—A second self.
Amende.—Compensation; apology.
Ami du peuple.—Friend of the people.
Amicus curiæ.—A friend of the court.
Amor patriæ.—Love of country.
Amour propre.—Self-love; vanity.
Ancien régime.—Former administration; ancient order of things.
Anglicè.—In English.
Anguis in herbâ.—A snake in the grass.
Animis opibusque parati.—Ever ready with our lives and property.
Animo et fide.—By (or with) courage and faith.
Animo facto.—Really and truly.
Animus furandi.—Felonious intent.
Anno Domini.—In the year of our Lord.
Anno lucis.—In the year of light.
Anno mundi.—In the year of the world.
Annus mirabilis.—Year of wonders.
Ante bellum.—Before the war.
Ante lucem.—Before light.
Ante meridiem.—Before noon.
Aperçu.—A brief sketch of any subject.
Apropos (Fr. *à propos*).—To the purpose.
Aqua vitæ.—Water of life; brandy.
Arbiter elegantiarum.—Master of ceremonies; an umpire in matters of taste.
Arcana imperii.—State secrets.
Arcanum.—A secret.
Argumentum ad crumenam.—An argument to the purse.
Argumentum ad fidem.—An appeal to faith.
Argumentum ad hominem.—An argument to the person.
Argumentum ad ignorantiam.—An argument founded on an adversary's ignorance of facts.
Argumentum ad judicium.—An appeal to the common sense of mankind.
Argumentum ad populum.—An appeal to the people.
Argumentum ad verecundiam.—An argument to modesty.
Argumentum baculinum.—Club law.
Armiger.—One bearing arms; an esquire.
Arrière-pensée.—Mental reservation.
Ars est celare artem.—True art is to conceal art.
Assumpsit.—It is assumed or taken for granted.
Astra castra, Numen lumen.—The stars my camp, the Deity my light.
At spes non fracta.—But hope is not broken.
Au fait.—Well instructed; master of it.
Au fond.—To the bottom, or main point.
Au pied de la lettre.—Literally.
Au pis aller.—At the worst.
Au revoir.—Farewell.
Audi alteram partem.—Hear the other side.
Aura popularis.—The gale of popular favour.
Auri sacra fames.—The accursed thirst for gold.
Aut vincere aut mori.—Victory or death.
Auto-da-fé, Auto-de-fe.—An act of faith; burning of heretics.
Auxilium ab alto.—Help from on high.
Avant-coureur.—A forerunner.

Badinage.—Light or playful discourse.
Bagatelle.—A trifle.
Bas bleu.—A blue-stocking; a literary woman.
Bateau.—A long light boat.
Beau-idéal.—A model of ideal perfection.
Beau monde.—The fashionable world.
Bel esprit.—A brilliant mind.
Bella-donna.—The deadly nightshade; fair lady.
Belles-lettres.—Polite literature.
Bellum internecinum.—A war of extermination.
Bellum lethale.—A deadly war.

Benigno numine.—By the favour of Providence.

Ben trovato.—Well found; an ingenious solution.

Billet-doux.—A love-letter.

Bis dat qui citò dat.—He gives twice who gives promptly.

Bis peccare in bello non licet.—To blunder twice is not allowed in war.

Bis vincit, qui se vincit in victoriâ.—He conquers a second time, who controls himself in victory.

Bizarre.—Odd; fantastic.

Blasé.—Surfeited.

Bon gré mal gré.—Willing or unwilling.

Bon jour.—Good-day; good-morning.

Bon mot.—A witty saying; a jest; a quibble.

Bon soir.—Good-evening.

Bon ton.—High fashion; first-class society.

Bon vivant.—A high liver.

Bona fide.—In good faith.

Bon-bon.—A sweetmeat; confectionery.

Bonhomie.—Good-natured simplicity.

Bonis nocet quisquis pepercerit malis.—He hurts the good who spares the bad.

Bonne bouche.—A delicious morsel.

Bonus.—An extra payment for a service rendered or a thing received.

Boreas.—The north wind.

Boudoir.—A small private apartment.

Bourgeois.—A citizen of the trading class; a printing type.

Bourgeoisie.—The body of citizens.

Bravura.—A song of difficult execution.

Breveté.—Patented.

Brutum fulmen.—A harmless thunderbolt; unreasoning bluster.

Burletta.—A musical farce

Cachet.—A seal.

Cacoethes.—A bad habit or custom.

Cacoethes carpendi.—A rage for finding fault.

Cacoethes loquendi.—An itch for speaking.

Cacoethes scribendi.—A passion for writing.

Cadenza.—The fall or modulation of the voice, in music.

Cæca est invidia.—Envy is blind.

Cætera desunt.—The remainder is wanting.

Cæteris paribus.—Other things being equal.

Calibre.—Capacity or compass; mental power; a term in gunnery.

Camera obscura.—A dark chamber used by artists.

Campus Martius.—The field of Mars; a place of military exercise.

Canaille.—The rabble.

Candida Pax.—White-robed Peace.

Cantata.—A poem set to music.

Cantate Domino.—Sing to the Lord.

Cap-à-pie.—From head to foot.

Capias ad satisfaciendum.—You may take to satisfy.

Capriccio.—A fanciful irregular kind of musical composition.

Capriole.—A leap without advancing; capers.

Caput mortuum.—Dead head; the worthless remains.

Caret.—Is wanting or omitted.

Caret initio et fine.—It wants beginning and end.

Carpe diem.—Enjoy the present day.

Carte blanche.—Unconditional terms.

Casus belli.—An occasion for war.

Casus fœderis.—A case of conspiracy; the end of the league.

Catalogue raisonné.—A catalogue of books arranged according to their subjects.

Cause célèbre.—A remarkable trial in a court of justice.

Caveat actor.—Let the doer beware.

Caveat emptor.—Let the purchaser take heed or beware.

Cavendo tutus.—Safe through caution.

Ce n'est que le premier pas qui coûte.—It is only the first step which is difficult.

Cedant arma togæ.—Let military power yield to the civil.

Cede Deo.—Submit to Providence.

Certiorari.—To be made more certain.

Cessio bonorum.—Yielding up of goods.

C'est une autre chose.—That is quite a different thing.

Chacun à son goût.—Every one to his taste.

Chanson.—A song.

Chansonnette.—A little song.

Chapeau.—A hat.

Chapelle ardente.—The place where a dead person lies in state.

Chaperon.—An attendant on a lady, as a guide and protector.

Chargé d'affaires.—An ambassador of second rank.

Château.—A castle; a country mansion.

Chef-d'œuvre.—A masterpiece.

Chevalier d'industrie.—A knight of industry; one who lives by persevering fraud.

Chi tace confessa.—Silence is confession.

Chiaro-oscuro or *Chiaroscuro.*—Light and shadow in painting.

Chose qui plaît est à demi vendue.—A thing which pleases is already half sold.

Cicerone.—A guide or conductor.

Cicisbeo.—A dangler after a lady

Ci-devant.—Formerly; former.

Citò maturum, citò putridum.—Soon ripe, soon rotten.

Clarior e tenebris.—More bright from obscurity.

Clique.—A party; a gang.

Cognomen.—A surname.

Comme il faut.—As it should be.

Commune bonum.—A common good.

Communia propriè dicere.—To express common things with propriety.

Communibus annis.—One year with another.

Compos mentis.—Of sound mind.

Con amore.—With love or hearty inclination.

Concio ad clerum.—A discourse to the clergy.

Congé d'élire.—Permission to elect.

Connoisseur.—A skilful judge.

Consensus facit legem.—Consent makes the law.

Contour.—The outline of a figure.

Contra.—Against.

Contra bonos mores.—Against good manners.

Contretemps.—A mischance; disappointment.

Coram nobis.—Before us.

Coram non judice.—Before one who is not the proper judge.

Cornucopia.—The horn of plenty.

Corpus delicti.—The whole nature of the offence.

Corrigenda.—Corrections to be made.

Coryphæus.—A leader, or chief.

Cotillon.—A lively dance.

Couleur de rose.—Rose-colour; an aspect of beauty and attractiveness.

Coup de grâce.—The finishing stroke.

Coup de main.—A bold and rapid enterprise.

Coup de pied.—A kick.

Coup de soleil.—A stroke of the sun.

Coup d'état.—A master-stroke of state policy.

Coup d'œil.—Rapid view or glance.

Coûte qu'il coûte.—Cost what it may.

Credat Judæus.—A Jew may believe it.

Crescit amor nummi quantum ipsa pecunia crescit.—The love of money increases as rapidly as the money itself increases.

Crescit eundo.—It increases by going.

Crescite et multiplicamini.—Increase and multiply.

Crimen falsi.—Falsehood; perjury.

Crux criticorum.—The cross or puzzle of critics.

Cui bono?—To what good or advantage?

Cui malo?—To what evil will it tend?

Cul de sac.—The bottom of the bag; a difficulty; a street or lane that has no outlet.

Cum grano salis.—With a grain of salt; with some allowance.

Cum multis aliis.—With many others.

Cum privilegio.—With privilege.

Curia advisari vult.—The court wishes to be advised.

Curiosa felicitas.—A felicitous tact.

Currente calamo.—With a running pen; written off-hand.

Custos rotulorum.—Keeper of the rolls.

Da capo.—Over again.

Damnant quod non intelligunt.—They condemn what they do not comprehend.

Data.—Things granted (sing. *datum*).

De bonis non.—Of the goods not yet administered on.

De die in diem.—From day to day.

De facto.—In fact; in reality.

De gustibus non est disputandum.—There is no disputing about tastes.

De jure.—By law or right.

De mortuis nil nisi bonum.—Say nothing but what is good of the dead.

De novo.—Anew.

De profundis.—Out of the depths.

De trop.—Out of place; not wanted.
Debito justitiæ.—By debt of justice.
Début.—Beginning of an enterprise; first appearance.
Deceptio visûs.—An illusion of the sight.
Dedimus potestatem.—We have given power.
Deficit.—A want or deficiency.
Dei gratiâ.—By the grace of God.
Déjeûner à la fourchette.—A breakfast or luncheon with meats.
Dele.—Blot out or erase.
Delenda est Carthago.—Carthage must be blotted out.
Delta (the Greek letter Δ), a triangular tract of land toward the mouth of a river.
Dénouement.—An unravelling or winding up.
Deo adjuvante, non timendum.—God helping, nothing need be feared.
Deo favente.—With God's favour.
Deo gratias.—Thanks to God.
Deo juvante.—With God's help.
Dêo, non fortunâ.—From God, not fortune.
Deo volente, or *D. V.*—God willing.
Dépôt.—A store; the recruiting reserve of regiments.
Dernier ressort.—The last resort.
Desideratum.—Something desired or wanted.
Desunt cætera.—The other things are wanting.
Détour.—A circuitous march.
Detur digniori.—Let it be given to the more worthy.
Deus ex machinâ.—A god from the clouds; unexpected aid in an emergency.
Devoir.—Duty.
Dexter.—The right hand.
Dictum.—A positive assertion (pl. *dicta*).
Dictum de dicto.—Report upon hearsay.
Dies faustus.—A lucky day.
Dies iræ.—Day of wrath.
Dies non.—A day on which judges do not sit.
Dieu et mon droit.—God and my right.
Dieu vous garde.—God protect you.
Dii majorum gentium.—The gods of the superior class; the twelve superior gods.
Dii penates.—Household gods.
Dilettanti.—Persons who devote themselves to science merely for amusement or relaxation. (Sing. *Dilettante.*)
Diluvium.—A deposit of superficial loam, sand, &c. caused by a deluge.
Dirigo.—I direct or guide.
Disjecta membra.—Scattered parts, limbs, or writings.
Distrait.—Absent in thought; absent-minded.
Distringas.—A writ for distraining.
Divide et impera.—Divide and govern.
Doce ut discas.—Teach, that you may learn.
Docendo dicimus.—We learn by teaching.
Dolce.—Soft and agreeable. (*Music.*)
Dolce far niente.—Sweet nothing-to-do.
Doli incapax.—Incapable of mischief.
Doloroso.—Soft and pathetic. (*Music.*)
Domicile (L. *domicilium*).—An abode.
Domine dirige nos.—O Lord, direct us.
Dominus vobiscum.—The Lord be with you.
Double entendre.—Double meaning (correctly written *double entente*).
Douceur.—A present or bribe; sweetness.
Draco.—A dragon; a constellation.
Dramatis personæ.—The characters in a play.
Duet (Ital. *duetto*).—A song for two performers.
Dulce est desipere in loco.—It is pleasant to jest, or revel, at the proper time.
Dulce et decorum est pro patriâ mori.—It is sweet and pleasant to die for one's country.
Dulia.—An inferior kind of worship.
Dum spiro, spero.—Whilst I breathe, I hope.
Dum vivimus, vivamus.—While we live, let us live.
Duo.—Two; a two-part song.
Duodecimo.—A book having twelve leaves to a sheet.
Durante placito, or *durante beneplacito.*—During pleasure.
Durante vitâ.—During life.
Dux fœmina facti.—A woman was the leader to the deed.

E pluribus unum.—One out of many; one composed of many: the motto of the United States.

Eau de vie.—Brandy; water of life.
Ecce homo.—Behold the man.
Ecce signum.—Behold the sign.
Eclaircissement.—The clearing-up of an affair.
Eclat.—Splendour; applause.
Editio princeps.—The first edition.
Elan.—Buoyancy; dash.
Elegit.—He hath elected; a writ of execution.
Elève.—A pupil.
Elite.—The best part.
Embonpoint.—Roundness; good condition.
Emeritus.—One retired from active official duties.
Emeute.—Insurrection; uproar.
Empressement.—Eagerness; ardour.
En ami.—As a friend.
En avant!—Forward!
En flûte.—Carrying guns on the upper deck only.
En grande tenue.—In full dress.
En masse.—In a mass; in a body.
En passant.—By the way; in passing.
En rapport.—In communication
En revanche.—In return.
En route.—On the way.
Enfans perdus.—Lost children; the forlorn hope.
Ennui.—Weariness; lassitude.
Ense petit placidam sub libertate quietem.—By his sword he seeks the calm repose of liberty.
Ensemble.—The whole taken together.
Entente cordiale.—The cordial understanding between two countries.
Entre nous.—Between ourselves.
Entrée.—Entrance.
Entremets.—Small and dainty dishes set between the principal ones at table.
Eo nomine.—By that name.
Equilibrium.—Equality of weight; even balance.
Ergo.—Therefore.
Eripuit cœlo fulmen, sceptrumque tyrannis.—He snatched the thunderbolt from heaven, and the sceptre from tyrants.
Erratum.—A mistake or error (pl. *errata*).
Espièglerie.—Waggish tricks.
Esprit de corps.—The animating spirit of a collective body.
Est modus in rebus.—There is a medium in all things.
Esto perpetua.—May it last forever.
Et cætera.—And the rest.
Eureka.—I have found it.
Ex.—Out of; late (as, ex-consul).
Ex animo.—Heartily.
Ex cathedrâ.—From the chair; with high authority.
Ex concesso.—From what has been granted.
Ex curiâ.—Out of court.
Ex fumo dare lucem.—Out of smoke to bring light.
Ex nihilo nihil fit.—Nothing can come of nothing.
Ex officio.—By virtue of his office.
Ex parte.—On one side only (before a noun, *exparte*).
Ex pede Herculem.—We recognize a Hercules from the size of the foot; that is, we judge of the whole from the specimen.
Ex post facto.—After the deed is done.
Ex tempore.—Without premeditation.
Ex uno disce omnes.—From one learn all; from one judge of the whole.
Excelsior.—More elevated; onward.
Excerpta.—Extracts.
Exempli gratiâ.—As for example.
Exeunt omnes.—All retire.
Experimentum crucis.—A decisive experiment.
Experto credo.—Believe one who has experience.
Exposé.—An exposition; recital.

Faber suæ fortunæ.—The architect of his own fortune.
Facile primus, facile princeps.—By far the first or chiefest.
Facilis est descensus.—Descent is easy.
Fac simile.—Make it like: hence, an exact copy.
Fac totum.—Do all: a man of all work.
Facta est lux.—There was light.
Fas est ab hoste doceri.—It is allowable to learn even from an enemy.
Fata obstant.—The fates oppose it.
Fauteuil.—An easy-chair.
Faux pas.—A false step.
Felo de se.—A self-murderer.
Feme couverte.—A married woman.
Feme sole.—A woman unmarried.

Festina lentè.—Hasten slowly; advance steadily rather than hurriedly.

Fête.—A feast or celebration.

Fête champêtre.—A rural feast.

Feu de joie.—A bonfire; a discharge of musketry on days of rejoicing.

Feuilleton.—A small leaf; a supplement to a newspaper; a pamphlet.

Fiat.—Let it be done.

Fiat justitia, ruat cœlum.—Let justice be done, though the heavens should fall.

Fiat lux.—Let there be light.

Fide, non armis.—By faith, not by arms.

Fide, sed cui vide.—Trust, but see whom.

Fides et justitia.—Fidelity and justice.

Fidus Achates.—Faithful Achates (that is, a true friend).

Fieri facias.—Cause it to be done (a kind of writ).

Filius nullius.—A son of nobody.

Fille-de-chambre.—A chambermaid.

Finale.—The close or end.

Finem respice.—Look to the end.

Finis.—The end.

Finis coronat opus.—The end crowns the work.

Flagrante bello.—While the war is raging.

Flagrante delicto.—In the commission of the crime.

Flâneur.—A lounger.

Flecti, non frangi.—To be bent, not to be broken.

Fleur-de-lis.—The flower of the lily (pl. *fleurs-de-lis*).

Forte.—In music, a direction to sing or play with force or spirit.

Fortes fortuna juvat.—Fortune assists the brave.

Fortissimo.—Very loud.

Fortiter in re.—Resolute in deed.

Fracas.—Bustle; a slight quarrel; more ado about the thing than it is worth.

Fruges consumere nati.—Born merely to consume the fruits of the earth.

Fugam fecit.—He has taken to flight.

Fuit Ilium.—Troy *has* been.

Functus officio.—Out of office.

Furore.—Excitement.

Gaieté de cœur.—Gayety of heart.

Gallicè.—In French.

Gardez bien.—Take good care.

Gardez la foi.—Keep the faith.

Gaucherie.—Awkwardness.

Gaudeamus igitur.—So let us be joyful.

Gendarme.—A military policeman.

Gendarmerie.—The body of the *gendarmes.*

Genius loci.—The genius of the place.

Genus irritabile vatum.—Irritable tribe of poets.

Gloria in excelsis.—Glory to God in the highest.

Gratis.—Free of cost.

Gratis dictum.—Mere assertion.

Gravamen.—The thing complained of.

Grisette.—Dressed in gray (a term applied to French shop-girls, &c.).

Gusto.—Great relish.

Habeas corpus.—You are to have the body: a writ of right, by virtue of which every citizen can, when imprisoned, demand to be put on his trial.

Habitué.—A frequenter.

Hæc olim meminisse juvabit.—It will be pleasant hereafter to remember these things.

Haricot.—A kind of ragout; a kidney-bean.

Haud passibus æquis.—Not with equal steps. [Wrongly quoted: see *Non*, &c.]

Haut goût.—High flavour.

Hauteur.—Haughtiness.

Helluo librorum.—A book-worm.

Hic et ubique.—Here, there, and everywhere.

Hic jacet.—Here lies.

Hinc illæ lacrymæ.—Hence proceed these tears.

Hoc age.—Do this; attend to what you are doing.

Homme d'esprit.—A man of talent, or of wit.

Homo multarum literarum.—A man of much learning.

Honi soit qui mal y pense.—Evil be to him that evil thinks.

Honores mutant mores.—Honours change men's manners.

Hora fugit.—The hour or time flies.

Horresco referens.—I shudder to relate

Hors de combat.—Disabled for fighting; vanquished.

Hortus siccus.—A collection of dried plants.
Hostis humani generis.—An enemy of the human race.
Hotel de ville.—A town-hall.
Hôtel-Dieu.—The chief hospital in French cities.
Humanum est errare.—It is human to err.
Hunc tu caveto.—Beware of him.

Ibidem, contracted *ibid.* or *id.*—In the same place.
Ich dien.—I serve.
Id est.—That is; abridged *i.e.*
Id genus omne.—All of that sort.
Idem, contracted *id.*—The same. (*Id. ib.*, the same author; in the same place.)
Idoneus homo.—A fit man.
Ignorantia legis neminem excusat.—Ignorance of the law excuses no one.
Il a le diable au corps.—The devil is in him.
Imitatores, servum pecus.—Imitators, a servile herd.
Imperium in imperio.—One government existing within another.
Impransus.—One who has not dined.
Imprimatur.—Let it be printed.
Imprimis.—In the first place.
Impromptu.—A prompt remark without study.
In articulo mortis.—At the point of death.
In capite.—In the head.
In cœlo quies.—There is rest in heaven.
In commendam.—In trust.
In conspectu fori.—In the eye of the law; in the sight of the court.
In curiâ.—In the court.
In duplo.—Twice as much.
In equilibrio.—Equally balanced.
In esse.—In being.
In extenso.—At full length.
In extremis.—At the point of death.
In formâ pauperis.—As a pauper.
In foro conscientiæ.—Before the tribunal of conscience.
In hoc signo vinces.—In this sign thou shalt conquer.
In limine.—At the threshold.
In loco.—In the place.
In medias res.—Into the midst of things.
In memoriam.—To the memory of.
In perpetuum.—Forever.
In petto.—In reserve; in one's breast.
In posse.—In possible existence.
In posterum.—For the time to come.
In propriâ personâ.—In his person.
In puris naturalibus.—Quite naked.
In re.—In the matter of.
In situ.—In its original situation.
In statu quo.—In the former state.
In te, Domine, speravi.—In thee, Lord, have I put my trust.
In terrorem.—By way of warning.
In totidem verbis.—In so many words.
In toto.—Altogether.
In transitu.—On the passage.
In utrumque paratus.—Prepared for either event.
In vacuo.—In empty space, or in a vacuum.
In vino veritas.—There is truth in wine.
Incognito.—Disguised; unknown.
Index expurgatorius.—A list of prohibited books.
Infra dignitatem.—Beneath one's dignity.
Innuendo.—Covert meaning; indirect hint.
Inops consilii.—Without counsel.
Insouciance.—Carelessness; indifference.
Instar omnium.—One will suffice for all; an example to others.
Inter alia.—Among other things.
Inter arma leges silent.—In the midst of arms the laws are silent.
Inter nos.—Between ourselves.
Inter se.—Among themselves.
Ipse dixit.—He himself said it; dogmatic assertion.
Ipsissima verba.—The very words.
Ipso facto.—By the fact itself; actually.
Ipso jure.—By the law itself.
Ira furor brevis est.—Anger is brief madness.
Ita lex scripta est.—Thus the law is written.
Item.—Also.

Jacta est alea.—The die is cast.
Jamais arrière.—Never behind.
Je ne suis quoi.—I know not what.
Jet d'eau.—A jet of water.
Jeu de mots.—Play upon words; a pun.
Jeu d'esprit.—A witticism.

Judicium Dei.—The judgment of God.
Juniores ad labores.—Young men for labours.
Jure divino.—By divine law.
Jure humano.—By human law.
Jus civile.—Civil law.
Jus gentium.—The law of nations.
Jus gladii.—Right of the sword.
Juste milieu.—The golden mean; a just medium.
Justitiæ soror fides.—Faith is the sister of justice.

La critique est aisée, et l'art est difficile.—Criticism is easy, but art is difficult.
Labor ipse voluptas.—Labour itself is pleasure.
Labor omnia vincit.—Labour conquers all things.
Laissez-nous faire.—Let us alone.
Lapsus calami.—A slip of the pen; an error in writing.
Lapsus linguæ.—A slip of the tongue.
Lapsus memoriæ.—A slip of memory.
Lares et penâtes.—Household gods.
L'argent.—Money, or silver.
Laudator temporis acti.—A praiser of time past.
Laus Deo.—Praise to God.
Laus propria sordet.—Praise of one's own self defiles.
Le beau monde.—The fashionable world.
Le bon temps viendra.—The good time will come.
Le grand œuvre.—The great work; the philosopher's stone.
Le pas.—Precedence in place or rank.
Le savoir-faire.—The knowledge how to act; address.
Le tout ensemble.—All together.
Lege.—Read.
Leges legum.—The law of laws.
Lèse majesté.—High treason.
L'étoile du nord.—The star of the north.
Lettre de cachet.—A sealed letter; a royal warrant.
Levée.—A morning visit or reception.
Lex loci.—The law of the place.
Lex magna est, et prævalebit.—The law is great, and will prevail.
Lex non scripta.—The unwritten or common law.
Lex scripta.—Statute law.
Lex talionis.—The law of retaliation.
Lex terræ, lex patriæ.—The law of the land.
L'homme propose, et Dieu dispose.—Man proposes, and God disposes.
Libretto.—A little book or pamphlet.
Licentia vatûm.—A poetical license.
Lingua Franca.—The mixed language spoken by Europeans in the East.
Liqueur.—A cordial.
Lis litem generat.—Strife begets strife.
Lis sub judice.—A case not yet decided.
Lite pendente.—During the trial.
Litera scripta manet.—The written letter remains.
Literati.—Men of letters or learning.
Loco citato.—In the place cited.
Locum tenens.—One who holds a place for another.
Locus sigilli (L. S.).—The place of the seal.
Longo intervallo.—At a great distance.
Ludere cum sacris.—To trifle with sacred things.
Lusus naturæ.—A sport or freak of nature.

Macte virtute.—Proceed in virtue.
Mademoiselle.—A young unmarried lady.
Magna Charta.—The great charter of England.
Magna civitas, magna solitudo.—A great city is a great desert.
Magna est veritas, et prævalebit.—The truth is great, and will prevail.
Magni nominis umbra.—The shadow of a great name.
Magnum opus.—A great work.
Magnus Apollo.—Great Apollo; one of high authority.
Maison de ville.—The town-house.
Maître d'hôtel.—An hotel-keeper; a house-steward.
Majordomo (Ital. *maiordomo*).—One who has the management of a household.
Malâ fide.—In bad faith; treacherously.
Mal à propos.—Out of time; unbecoming.
Malaria.—Noxious exhalations.
Malgré.—In spite of.
Malum in se.—Bad in itself.
Mandamus.—We command: a writ from the Queen's Bench.
Manège.—A riding-school.

Mania a potu.—Madness caused by drunkenness.
Manu forti.—With a strong hand.
Mardi gras.—Shrove-Tuesday.
Mare clausum.—A closed sea; a bay.
Materfamilias.—The mother of a family.
Materia medica.—Substances used in the healing art.
Matinée.—A morning party.
Mauvais goût.—Bad taste.
Mauvais sujet.—A worthless fellow.
Mauvaise honte.—False modesty; bashfulness.
Maximum.—The greatest.
Maximus in minimis.—Very great in trifling things.
Me judice.—I being judge; in my own opinion.
Medio tutissimus ibis.—A medium course will be safest.
Meditatione fugæ.—In contemplation of flight.
Memento mori.—Remember death.
Memorabilia.—Things to be remembered.
Memoriter.—By rote.
Ménage.—Household.
Mens sana in corpore sano.—A sound mind in a sound body.
Mens sibi conscia recti.—A mind conscious of rectitude.
Mensa et thoro.—From bed and board.
Merum sal.—Pure salt; genuine Attic wit.
Meum et tuum.—Mine and thine.
Minimum.—The least.
Minutiæ.—Minute concerns; trifles.
Mirabile dictu.—Wonderful to be told.
Mirabilia.—Wonders.
Mittimus.—We send: a warrant for the commitment of an offender.
Modus operandi.—Manner of operation.
Montani semper liberi.—Mountaineers are always freemen.
Morceau.—A morsel.
More suo.—In his own way.
Mot du guet.—A watchword.
Multum in parvo.—Much in a small space.
Mutanda.—Things to be altered.
Mutatis mutandis.—The needful changes being made.
Mutato nomine.—The name being changed.
Naïveté.—Ingenuousness; simplicity.
Ne cede malis.—Yield not to misfortune.
Ne exeat.—Let him not depart.
Ne plus ultra.—Nothing further; the uttermost point.
Ne quid nimis.—Not too much of any thing; do nothing to excess.
Ne sutor ultra crepidam.—Let not the shoemaker go beyond his last.
Ne tentes, aut perfice.—Attempt not, or accomplish thoroughly.
Nec pluribus impar.—Not an unequal match for numbers.
Nec scire fas est omnia.—It is not permitted to know all things.
Necessitatis non habet legem.—Necessity has no law.
Née.—Born.
Nefasti dies.—Days upon which no public business was transacted; also, unlucky days.
Nemine contradicente.—No one contradicting.
Nemine dissentiente.—Without opposition or dissent.
Nemo me impune lacessit.—No one wounds me with impunity.
Nemo mortalium omnibus horis sapit.—No one is wise at all times.
Nemo repentè fuit turpissimus.—No man ever became a villain at once.
Nemo solus sapit.—No one is wise alone.
Niaiserie.—Silliness.
Nihil debet.—He owes nothing: a plea denying a debt.
Nihil quod tetigit, non ornavit.—Whatever he touched he embellished.
Nil admirari.—To wonder at nothing.
Nil desperandum.—Never despair.
Nimium ne crede colori.—Trust not too much to looks.
N'importe.—It matters not.
Nisi Dominus frustra.—Unless the Lord be with us, all efforts are in vain.
Noblesse oblige.—Rank imposes obligation.
Nolens volens.—Willing or unwilling.
Noli me tangere.—Don't touch me.
Nolle prosequi.—Unwilling to proceed.
Nolo episcopari.—I am not willing to be made a bishop (an old formal way of declining a bishopric).
Nom de guerre.—An assumed name.
Nom de plume.—A literary title.

Nomen et omen.—Name and omen; a name that is ominous.

Non compos mentis.—Not of sound mind.

Non deficiente crumenâ.—If the money does not fail.

Non est disputandum.—It is not to be disputed.

Non est inventus.—Not found.

Non libet.—It does not please me.

Non mi ricordo.—I don't remember.

Non nobis solum.—Not merely for ourselves.

Non obstante.—Notwithstanding.

Non omnis moriar.—I shall not wholly die.

Non passibus æquis.—Not with equal steps.

Non sequitur.—It does not follow: an unwarranted conclusion.

Non sibi, sed omnibus.—Not for itself, but for all.

Nonchalance.—Coolness; easy indifference.

Nonpareil.—Peerless; a small printing type.

Nosce teipsum.—Know thyself.

Noscitur ex sociis.—He is known by his companions.

Nota bene.—Mark well.

Nous verrons.—We shall see.

Novus homo.—A new man.

Nudum pactum.—An invalid agreement.

Nulla nuova, bona nuova.—The best news is no news.

Nullius filius.—The son of nobody.

Nunc aut nunquam.—Now or never.

O tempora! o mores!—Oh, the times! oh, the manners!

Obiit.—He (or she) died.

Obiter dictum.—A thing said by the way, or in passing.

Obsta principiis.—Resist the first beginnings.

Odi profanum.—I loathe the profane.

Odium theologicum.—The hatred of theologians.

Ohe! jam satis.—Oh, there is now enough.

Olla podrida.—An incongruous mixture.

Omne ignotum pro magnifico.—Whatever is unknown is thought to be magnificent.

Omnes.—All.

Omnia bona bonis.—All things are good with the good.

Omnia vincit amor.—Love conquers all things.

On-dit.—A rumour; a flying report.

Onus.—Burden.

Onus probandi.—The responsibility of producing proof.

Ope et consilio.—With assistance and counsel.

Ora et labora.—Pray and work.

Orator fit, poeta nascitur.—The orator is made by education, but a poet must be born.

Ore rotundo.—With full-sounding voice.

Otium cum dignitate.—Dignified leisure.

Outré.—Preposterous; eccentric.

Pallida mors.—Pale death.

Par excellence.—By way of eminence.

Par nobile fratrum.—A noble pair of brothers; two just alike.

Pari passu.—With equal step; in the same degree.

Parole d'honneur.—Word of honour.

Pars pro toto.—Part for the whole.

Particeps criminis.—An accomplice.

Parturiunt montes, nascetur ridiculus mus.—The mountains are in labour; a ridiculous mouse will be brought forth.

Parva componere magnis.—To compare small things with great.

Parvenu.—A new-comer; an upstart.

Pas.—A step; precedence.

Passe-partout.—A master-key.

Passim.—In many places; everywhere.

Paterfamilias.—The father of a family.

Pater noster.—Our Father; the Lord's prayer.

Pater patriæ.—Father of his country.

Patois.—A corrupt dialect.

Pax in bello.—Peace in war.

Peccavi.—I have sinned.

Penchant.—An inclination; a leaning toward.

Pendente lite.—While the suit is pending.

Penetralia.—Secret recesses.

Per aspera ad astra.—Through trials to glory.

Per capita.—By the head.

Per cent. or *per centum.*—By the hundred

Per contra.—Contrariwise.
Per curiam.—By the court.
Per diem.—By the day.
Per fas et nefas.—Through right and wrong.
Per saltum.—With a leap; at once.
Per se.—By itself; alone.
Perdu.—Concealed.
Père de famille.—The father of a family.
Petit.—Small; little.
Petitio principii.—A begging of the question.
Petit-maître.—A fop.
Peu à peu.—Gradually; little by little.
Pinxit.—Painted it: placed after the artist's name on a picture.
Più.—More.
Plateau.—A plain; a flat surface.
Plebs.—Common people.
Poco.—A little.
Poeta nascitur, non fit.—A poet is born, not made.
Point d'appui.—Point of support; prop.
Poisson d'Avril.—April fool.
Populus vult decipi.—People like to be deceived.
Posse comitatûs.—The power of the county.
Post mortem.—After death.
Postulata.—Things assumed.
Præcognita.—Things previously known.
Præmonitus, præmunitus.—Forewarned, forearmed.
Preux chevalier.—A brave knight.
Primâ facie.—On the first view.
Primum mobile.—The primary motive, or moving power.
Primus inter pares.—Chief among equals.
Principia, non homines.—Principles, not men.
Principiis obsta.—Resist the first innovations.
Pro aris et focis.—For our altars and our hearths.
Pro bono publico.—For the public good.
Pro et con (for *contra*).—For and against.
Pro formâ.—For form's sake; according to form.
Pro hâc vice.—For this turn or occasion.
Pro loco et tempore.—For the place and time.
Pro patriâ.—For our country.
Pro ratâ.—In proportion.
Pro re natâ.—For a special emergency.
Pro tanto.—For so much.
Pro tempore.—For the time-being.
Probatum est.—It has been tried and proved.
Procès-verbal.—A written statement.
Procul, O procul este, profani!—Far, far hence, O ye profane!
Pronunciamento.—A public declaration.
Propagandâ fide.—For extending the faith.
Protégé.—A person taken charge of, or patronized; a ward, &c.
Prudens futuri.—Thoughtful of the future.
Pugnis et calcibus.—With fists and heels; with all the might.
Punica fides.—Punic faith; treachery.

Quære.—Query; inquiry.
Quamdiu se bene gesserit.—So long as he shall conduct himself properly.
Quantum.—The due proportion.
Quantum libet.—As much as you please.
Quantum meruit.—As much as he deserved.
Quantum sufficit.—A sufficient quantity; enough.
Quare impedit.—Why he hinders.
Quasi dicas.—As if you should say.
Quelque chose.—A trifle.
Qui capit, ille facit.—He who takes it makes it.
Qui pense?—Who thinks?
Qui tam?—Who as well? the title given to a certain action at law.
Qui transtulit sustinet.—He who brought us hither still preserves us.
Qui va là?—Who goes there?
Qui vive?—Who goes there? hence, on the *qui-vive*, on the alert.
Quid-nunc?—What now? a newsmonger.
Quid pro quo.—One thing for another; "tit for tat."
Quid rides?—Why do you laugh?
Quis separabit?—Who shall separate us?
Quo animo?—With what intention?
Quo jure?—By what right?
Quo warranto.—By what warrant or authority.
Quoad hoc.—To this extent.
Quod avertat Deus!—Which may God avert!

Quod vide.—Which see.
Quodlibet.—A nice point; a subtlety.
Quondam.—Former.
Quorum.—Of whom: a term signifying a sufficient number for a certain business.
Quos Deus vult perdere, prius dementat.—Those whom God wishes to destroy, he first deprives of understanding.

Ragout.—A highly-seasoned dish.
Rara avis.—A rare bird; a prodigy.
Re infectâ.—The business being unfinished.
Recte et suaviter.—Justly and mildly.
Rectus in curiâ.—Upright in the court; with clean hands.
Redolet lucernâ.—It smells of the lamp; it is a laboured production.
Reductio ad absurdum.—A reducing a position to an absurdity.
Regina.—Queen.
Regium donum.—A royal donation (a grant from the British crown to the Irish Presbyterian clergy).
Regnant populi.—The people rule.
Rencontre.—An encounter.
Requiescant in pace.—May they rest in peace.
Requiescat in pace.—May he rest in peace.
Rerum primordia.—The first elements of things.
Res angusta domi.—Narrow circumstances at home; poverty.
Respice finem.—Look to the end.
Respublica.—The commonwealth.
Restaurateur.—A tavern-keeper who provides dinners, &c.
Résumé.—An abstract or summary.
Resurgam.—I shall rise again.
Revenons à nos moutons.—Let us return to our subject.
Rex.—King.
Rouge.—Red colouring for the skin.
Rouge et noir.—Red and black (a kind of game).
Rus in urbe.—The country in town.
Ruse contre ruse.—Diamond cut diamond; trick for trick.
Ruse de guerre.—A stratagem of war.

Salle.—Hall.
Salus populi suprema lex est.—The welfare of the people is the supreme law
Salvo pudore.—Without offence to modesty.
Sanctum sanctorum.—Holy of Holies
Sang-froid.—Coolness; self-possession.
Sans.—Without.
Sans cérémonie.—Without ceremony.
Sans peur et sans reproche.—Without fear and without reproach.
Sans souci.—Without care; free and easy.
Sans tâche.—Stainless.
Sans-culottes.—Without breeches: a term applied to the rabble of the French Revolution.
Sartor resartus.—The cobbler mended.
Satis, superque.—Enough, and more than enough.
Satis verborum.—Enough of words; you need say no more.
Sauve qui peut.—Save himself who can.
Savant.—A learned man.
Savoir-faire.—Ability; skill.
Scandalum magnatum.—Scandal of the great.
Scienter.—Knowingly.
Scilicet.—That is to say; to wit.
Scire facias.—Cause it to be known.
Scripsit.—Wrote it.
Sculpsit.—Engraved it: placed after the engraver's name in prints.
Secundum artem.—According to rule.
Selon les règles.—According to rule.
Semper fidelis.—Always faithful.
Semper idem.—Always the same.
Semper paratus.—Always ready.
Senatûs consultum.—A decree of the senate.
Seriatim.—In order; successively.
Si quæris peninsulam amœnam, circumspice.—If thou seekest a beautiful peninsula, behold it here.
Sic in originali.—So it stands in the original.
Sic itur ad astra.—Such is the way to immortality.
Sic passim.—So everywhere.
Sic semper tyrannis.—So be it ever to tyrants.
Sic transit gloria mundi.—Thus passes away the glory of the world.
Sicut ante.—As before.
Similia similibus curantur.—Like things are cured by like.
Simplex munditiis.—Of simple elegance.

Sine die.—Without naming a day.
Sine invidiâ.—Without envy.
Sine qua non.—An indispensable requisite.
Siste, viator.—Stop, traveller.
Sobriquet.—A nickname.
Soi-disant.—Self-styled; pretended.
Soirée.—An evening party.
Souvenir.—Remembrance; a keepsake.
Spectas et spectaberis.—You will see and be seen.
Spes mea Christus.—Christ is my hope.
Spolia opima.—The richest body.
Stans pede in uno.—Standing on one foot.
Statu quo, or *in statu quo.*—In the same state.
Stet.—Let it stand.
Suaviter in modo, fortiter in re.—Gentle in manner, resolute in deed.
Sub judice.—Under consideration.
Sub rosâ.—Under the rose; privately.
Sub silentio.—In silence.
Subpœna.—Under a penalty: a summons to attend a court as a witness.
Succedaneum.—A substitute.
Sui generis.—Of its own kind; peculiar.
Summum bonum.—The chief good.
Super visum corporis.—Upon a view of the body.
Suppressio veri, suggestio falsi.—A suppression of the truth is the suggestion of a falsehood.
Supra.—Above.
Suum cuique.—Let every one have his own.

Table d'hôte.—An ordinary at which the master of the hotel presides.
Tabula rasa.—A smooth or blank tablet.
Tædium vitæ.—Weariness of life.
Tale quale.—Such as it is.
Tant mieux.—So much the better.
Tant pis.—So much the worse.
Tapis.—The carpet.
Tartuffe.—A nickname for a hypocritical devotee, derived from the principal character in Molière's comedy so called.
Te judice.—You may judge.
Tempora mutantur, et nos mutamur in illis.—The times are changed, and we are changed with them.
Tempus edax rerum.—Time the devourer of all things.
Tempus fugit.—Time flies.
Tempus omnia revelat.—Time reveals all things.
Teres atque rotundus.—Smooth and round; polished and complete.
Terra firma.—Solid earth; a safe footing.
Terra incognita.—An unknown country.
Tertium quid.—A third something; a nondescript.
Tête-à-tête.—A conversation between two parties.
Tirade.—A tedious and bitter harangue.
Ton.—The fashion.
Torso.—The fragmentary trunk of a statue.
Tot homines, quot sententiæ.—So many men, so many minds.
Totidem verbis.—In just so many words.
Toties quoties.—As often as.
Toto cœlo.—By the whole heavens; diametrically opposite.
Toto corde.—With the whole heart.
Toujours prêt.—Always ready.
Tour à tour.—By turns.
Tout bien ou rien.—The whole or nothing.
Tout ensemble.—The whole.
Tria juncta in uno.—Three united in one.
Tu quoque, Brute!—And thou too, Brutus!
Tuebor.—I will defend.
Tutto è buono che vien da Dio.—All is good which comes from God.
Tuum est.—It is your own.

Ubi jus incertum, ibi jus nullum.—Where the law is uncertain, there is no law.
Ubi libertas, ibi patria.—Where liberty dwells, there is my country.
Ubi supra.—Where above mentioned.
Ultima ratio regum.—The last argument of kings; military weapons; war.
Ultima Thule.—The utmost boundary or limit.
Ultimatum.—A final answer or decision.
Un bel esprit.—A wit; a virtuoso.
Un sot à triple étage.—An egregious blockhead.
Unâ voce.—With one voice; unanimously.

Unique.—Singular; the only one of its kind.
Usque ad nauseam.—To disgust.
Usus loquendi.—Usage in speaking.
Ut infra.—As below.
Uti possidetis.—As you possess; state of present possession.
Utile dulci.—Utility with pleasure.

Vade-mecum.—Go with me; a constant companion.
Væ victus.—Woe to the vanquished!
Vale.—Farewell.
Valet-de-chambre.—A servant who assists his master in dressing.
Variæ lectiones.—Various readings.
Veluti in speculum.—As in a mirror.
Veni, vidi, vici.—I came, I saw, I conquered.
Verbatim et literatim.—Word for word and letter for letter.
Verbum sat sapienti.—A word is enough for a wise man.
Verdad es verde.—Truth is green.
Veritas vincit.—Truth conquers.
Versus.—Against; toward.
Vertu, Virtù.—Virtue; taste; art; skill.
Veto.—I forbid.
Vi et armis.—By force and arms.
Viâ.—By the way of.
Via media.—A middle course.
Vice.—In the room of.
Vice versâ.—The terms being exchanged; reversely.
Vide.—See.
Vide et crede.—See and believe.
Vide ut supra.—See as above.
Videlicet.—To wit; namely.
Videttes.—Sentinels on horseback.
Vignette.—A name given to slight engravings with which books, bank-notes, &c. are ornamented.
Vincit amor patriæ.—Love of country prevails.
Vinculum matrimonii.—The bond of marriage.
Virtuoso.—One skilled in matters of taste or art.
Virtute officii.—By virtue of office.
Vis inertiæ.—Inert power; the tendency of every body to remain at rest.
Vis medicatrix naturæ.—The healing tendency of nature.
Vis poetica.—Poetic genius.
Vis vitæ.—The vigour of life
Vis-à-vis.—Face to face.
Vita brevis, ars longa.—Life is short, and art is long.
Vivâ voce.—By word of mouth; by the living voice.
Vivant rex et regina.—Long live the king and queen.
Vivat regina.—Long live the queen.
Vivat respublica.—Live the republic.
Vive la bagatelle.—Success to trifling.
Vive la reine.—Long live the queen.
Vive l'empereur.—Long live the emperor.
Vive le roi.—Long live the king.
Vive l'impératrice.—Long live the empress.
Vive, vale.—Farewell, and be happy.
Voilà tout.—That's all.
Voilà une autre chose.—That's quite a different matter.
Volens et potens.—Willing and able.
Volgo gran bestia.—The mob is a great beast.
Vox, et præterea nihil.—A voice, and nothing more.
Vox populi, vox Dei.—The people's voice is God's voice.
Vox stellarum.—The voice of the stars: applied to almanacs.
Vulgò.—Vulgarly; commonly.
Vulnus immedicabile.—An irreparable injury.
Vultus est index animi.—The countenance is the index of the mind.

Zonam solvere.—To loose the virgin zone.

INDEX.

ET FACTA EST LUX.

ELECTROTYPED BY L. JOHNSON & CO.
PHILADELPHIA.
PRINTED BY JAMES B. RODGERS.

www.ingramcontent.com/pod-product-compliance
Lightning Source LLC
Chambersburg PA
CBHW031146120726
47905CB00006B/1830
9783337252083